Lecture Notes Editorial Policies

Lecture Notes in Statistics provides a format for the informal and quick publication of monographs, case studies, and workshops of theoretical or applied importance. Thus, in some instances, proofs may be merely outlined and results presented which will later be published in a different form.

Publication of the Lecture Notes is intended as a service to the international statistical community, in that a commercial publisher, Springer, can provide efficient distribution of documents that would otherwise have a restricted readership. Once published and copyrighted, they can be documented and discussed in the scientific literature.

Lecture Notes are reprinted photographically from the copy delivered in camera-ready form by the author or editor. Springer provides technical instructions for the preparation of manuscripts. Volumes should be no less than 100 pages and preferably no more than 400 pages. A subject index is expected for authored but not edited volumes. Proposals for volumes should be sent to one of the series editors or addressed to "Statistics Editor" at Springer in New York.

Authors of monographs receive 50 free copies of their book. Editors receive 50 free copies and are responsible for distributing them to contributors. Authors, editors, and contributors may purchase additional copies at the publisher's discount. No reprints of individual contributions will be supplied and no royalties are paid on Lecture Notes volumes. Springer secures the copyright for each volume.

Series Editors:

Professor P. Bickel
Department of Statistics
University of California
Berkeley, CA 94720
USA

Professor P. Diggle
Department of Mathematics
University of Lancaster
Lancaster LA1 4YF
England

Professor S. Fienberg
Department of Statistics
Carnegie-Mellon University
Pittsburgh, PA 15213
USA

Professor Dr. U. Gather
Department of Statistics
University of Dortmund
Dekanat
Vogelpothsweg 87
D-44221 Dortmund
Germany

Professor I. Olkin
Department of Statistics
Stanford University
Stanford, CA 94305
USA

Professor S. Zeger
Department of Biostatistics
The Johns Hopkins University
615 N. Wolfe Street
Baltimore, MD 21205-2103
USA

Lecture Notes in Statistics   **188**

Edited by P. Bickel, P. Diggle, S. Fienberg, U. Gather,
I. Olkin, S. Zeger

Constance van Eeden

Restricted Parameter Space Estimation Problems
Admissibility and Minimaxity Properties

 Springer

Constance van Eeden
Moerland 19
1151 BH Broek in Waterland
The Netherlands
cve@xs4all.nl

Library of Congress Control Number: 2006924372

ISBN-10: 0-387-33747-4
ISBN-13: 978-0387-33747-0

Printed on acid-free paper.

9 8 7 6 5 4 3 2 1

springer.com

Restricted-Parameter-Space
Estimation Problems
Admissibility and Minimaxity Properties

Constance van Eeden

To the memory of Jan Hemelrijk (1918–2005),

who, in the early 1950s, encouraged me to work

on restricted-parameter-spaces-inference problems

and was very supportive during the preparation

of my 1958 PhD thesis on the subject.

Preface

This monograph is addressed to anyone interested in the subject of restricted-parameter-space estimation, and in particular to those who want to learn, or bring their knowledge up to date, about (in)admissibility and minimaxity problems for such parameter spaces.

The coverage starts in the early 1950s when the subject of inference for restricted parameter spaces began to be studied and ends around the middle of 2004. It presents known, and also some new, results on (in)admissibility and minimaxity for nonsequential point estimation problems in restricted finite-dimensional parameter spaces. Relationships between various results are discussed and open problems are pointed out. Few complete proofs are given, but outlines of proofs are often supplied. The reader is always referred to the published papers and often results are clarified by presenting examples of the kind of problems an author solves, or of problems that cannot be solved by a particular result.

The monograph does not touch on the subject of testing hypotheses in restricted parameter spaces. The latest books on that subject are by Robertson, Wright and Dykstra (1988) and Akkerboom (1990), but many new results in that area have been obtained since.

The monograph does have a chapter in which questions about the existence of maximum likelihood estimators are discussed. Some of their properties are also given there as well as some algorithms for computing them. Most of these results cannot be found in the Robertson, Wright, Dykstra book.

The author's long familiarity with the subject of this monograph combined with her 14-year General Editorship of Statistical Theory and Method Abstracts make it very unlikely that she missed any published results on the subject.

The author thanks her co-author and good friend Jim Zidek for proposing she write a monograph on estimation in restricted parameter spaces and for his encouragement and help during the writing.

A large part of the work on this monograph was done while the author was spending fall semesters at the Department of Statistics at The University of British Columbia. Many thanks for the support and the hospitality.

Also, many thanks to the librarians of CWI (Centrum for Wiskunde en Informatica, Amsterdam, The Netherlands) for their help in locating copies of needed publications.

Finally, my thanks to several anonymous reviewers for their helpful comments, as well as to the staff at Springer's editorial and production departments – in particular to John Kimmel, Statistics Editor, – for their help in getting this monograph in final form.

Broek in Waterland, June 2006

Contents

1

Introduction: Some history and some examples

Sometime in the early 1950s somebody came to the Statistical Consulting Service of the Mathematical Center in Amsterdam with a practical problem which led to the following theoretical question. Suppose $X_i \sim^{ind} \text{Bin}(n_i, \theta_i)$, $i = 1, 2$, and suppose that it is known that $\theta_1 \leq \theta_2$. How does one estimate $\theta = (\theta_1, \theta_2)$? The client was given the maximum likelihood estimator (MLE) $\hat{\theta} = (\hat{\theta}_1, \hat{\theta}_2)$, where $\hat{\theta}_i = X_i/n_i, i = 1, 2$ if $X_1/n_1 \leq X_2/n_2$ and $\hat{\theta}_1 = \hat{\theta}_2 = (X_1 + X_2)/(n_1 + n_2)$ if not.

This led to the study of the k-sample problem where $X_{i,j}, j = 1, \ldots, n_i, i = 1, \ldots, k, k \geq 2$, are independent random variables and, for $i = 1, \ldots, k$, the $X_{i,j}$ have distribution function $F_i(x; \theta_i), \theta_i \in R^1$. The parameter space Θ for $\theta = (\theta_1, \ldots, \theta_k)$ was determined by inequalities among the θ_i and bounds on them. The inequalities among them were either a simple ordering, i.e., $\Theta = \{\theta \mid \theta_1 \leq \ldots \leq \theta_k\}$, or an incomplete one such as, e.g., $k = 3$ and $\Theta = \{\theta \mid \theta_1 \leq \theta_2, \theta_1 \leq \theta_3\}$. Conditions for the existence of the MLE of θ were obtained as well as algorithms for finding MLEs (see van Eeden, 1956, 1957a – c, 1958). The MLE for the simply ordered binomial case was also obtained by Ayer, Brunk, Ewing, Reid and Silverman (1955). Further, Brunk (1955, 1958) considered the k-sample problem where the $X_{i,j}$ have a one-parameter exponential-family distribution.

Many of these and related results concerning MLEs for ordered parameters can be found in Barlow, Bartholomew, Bremner and Brunk (1972) and in Robertson, Wright and Dykstra (1988). These books do not discuss properties like admissibility or minimaxity. In fact, in the early days of the development of restricted-parameter-space inference there does not seem to have been much interest, if any, in the properties of the MLE, nor for that matter in looking for other, possibly better, estimators. It seems that it was not known that estimators which have "good" properties in unrestricted spaces lose many of these properties when the parameter space is restricted. As an example, for

$X \sim \mathcal{N}(\theta, 1)$ and squared error loss, the MLE of θ for the parameter space $\{\theta \mid -\infty < \theta < \infty\}$ is unbiased, admissible, minimax and has a normal distribution. For the parameter space $\{\theta \mid \theta \geq 0\}$ the MLE is biased and inadmissible (see Sacks, 1960, 1963), but still minimax (see Katz, 1961). It does not have a normal distribution and for $\theta = 0$ it is, for a sample $X_1, \ldots, X_n$, not even asymptotically normal. But, there are cases where the MLE does not lose its admissibility property when the parameter space is restricted. For example, when $X \sim \text{Bin}(n, \theta)$ with $\theta \in [0, m]$ for some known $m \in (0, 1)$ and $n \geq 2$, the MLE of θ is admissible for squared error loss when $0 < mn \leq 2$ (see Charras and van Eeden, 1991a).

Examples of other problems addressed in the restricted-parameter-space-estimation literature are: (i) finding (admissible) dominators for inadmissible estimators, (ii) finding "good" estimators (admissible minimax, e.g.) and (iii) when do "good" properties of an estimator of a vector parameter carry over to its components? All of these properties depend, of course, on the loss function used. Most authors use (scale-invariant) squared-error loss, but other loss functions are being used and questions of universal admissibility (in the sense of Hwang, 1985) are studied by some authors. Finally, problems with nuisance parameters can give interesting and curious results. In such cases, restrictions imposed among the nuisance parameters (or between the nuisance parameters and the parameters of interest) can lead to improved estimation of the parameters of interest.

As will be seen, not very much is known about the question of how much "better" (risk-function-wise) dominators are than their inadmissible counterparts. Nor is much known about the possible improvements which can be obtained by restricting a parameter space. The numerical results on these two questions indicate that gains in the minimax value from restricting the parameter space can be very substantial when the restricted space is bounded. However, for unbounded restricted parameter spaces, the minimax value for the restricted space is often equal to the one for the unrestricted one. As far as gains from dominators is concerned, any gain one finds for a particular problem is only a lower bound on possible gains for that problem, unless one can show that there is no "better" dominator. And the obtained results on these lower bounds are very model dependent.

In this monograph, known (as well as some new) results on the above-mentioned aspects of estimation in restricted parameter spaces are described and discussed for the case of non-sequential point estimation in R^k. Relationships between the results of various authors, as well as open problems, are pointed out. Essential errors are reported on.

A general statement of the problem, as well as the notation and some definitions, are given in Chapter 2. Chapters 3 and 4 contain, respectively, results on admissibility and minimaxity when the problem does not contain any nuisance parameters. Results for the case where nuisance parameters are present are presented in Chapter 5 and results for the linear model are given in Chapter 6. Several other properties of and questions about restricted-parameter-space estimators, such as, e.g., robustness to misspecification of the parameter space and unbiasedness can be found in Chapter 7. Also given in that chapter are relationships with Hu and Zidek's weighted likelihood estimation (see F. Hu, 1994, 1997 and Hu and Zidek, 2002).

The last chapter, Chapter 8, contains existence results for maximum likelihood estimators under order-restrictions on the parameters as well as some of their properties and some algorithms to compute them. It is hoped that these results will help the reader better understand some of the presented results concerning maximum likelihood estimators in restricted parameter spaces.

An extensive bibliography concludes the monograph.

2

A statement of the problem, the notation and some definitions

Consider a probability space $(\mathcal{X}, \mathcal{A})$, a family of distributions

$$\mathcal{P}_o = \{P_{\theta,\lambda}, \theta = (\theta_1, \ldots, \theta_M), \lambda = (\lambda_1, \ldots, \lambda_{K-M}), (\theta, \lambda) \in \Omega_o \subset R^K\}$$

defined over it and a random vector $X \in R^n$ defined on it, where $M \geq 1$, $K - M \geq 0$ and Ω_o is closed and convex with a non-empty interior. Assume that $P_{\theta,\lambda}$ has a density, $p_{\theta,\lambda}$, with respect to a σ-finite measure ν. Then the problem considered in this monograph is the estimation, based on X, of θ when it is known that $(\theta, \lambda) \in \Omega$, where Ω is a known closed, convex proper subset of Ω_o with a non-empty interior. When $K - M \geq 1$, λ is a vector of nuisance parameters.

Let

$$\left.\begin{array}{c} \Theta_o = \{\theta \in R^M \mid (\theta, \lambda) \in \Omega_o \text{ for some } \lambda \in R^{K-M}\} \\[2mm] \Theta = \{\theta \in R^M \mid (\theta, \lambda) \in \Omega \text{ for some } \lambda \in R^{K-M}\}. \end{array}\right\} \qquad (2.1)$$

Then the set Θ is a known closed, convex subset of Θ_o with a non-empty interior.

For a definition of what, in this monograph, is considered to be an estimator of θ and to see its relationship to definitions used by other authors, first look at the case where

$$\text{the support of } P_{\theta,\lambda} \text{ is independent of } (\theta, \lambda) \text{ for } (\theta, \lambda) \in \Omega_o. \qquad (2.2)$$

Then estimators δ of θ based on X satisfy

$$P_{\theta,\lambda}(\delta(X) \in \Theta) = 1 \text{ for all } (\theta, \lambda) \in \Omega. \qquad (2.3)$$

This class of estimators is denoted by

$$\mathcal{D} = \{\delta \mid (2.3) \text{ is satisfied }\}. \qquad (2.4)$$

As examples of this kind of model, let X_1 and X_2 be independent random variables with $X_1 \sim \mathcal{N}(\theta, 1)$ and $X_2 \sim \mathcal{N}(\lambda, 1)$. Suppose θ and λ are unknown, but it is known that $\theta \leq \lambda \leq 1$. Then $K = 2$, $M = 1$, $\Omega_o = R^2$, $\Omega = \{(\theta, \lambda) \mid \theta \leq \lambda \leq 1\}$, $\Theta_o = R^1$ and $\Theta = \{\theta \mid \theta \leq 1\}$. In this case the problem is the estimation of θ based on $X = (X_1, X_2)$ by an estimator $\delta(X)$ satisfying $P_{\theta,\lambda}(\delta(X) \leq 1) = 1$ for all $(\theta, \lambda) \in \Omega$. For the case where $X_i \sim^{ind} \mathcal{N}(\theta_i, 1)$, $i = 1, 2$, with θ_1 and θ_2 unknown and $\theta_1 \leq \theta_2$, $K = M = 2$ and $\theta = (\theta_1, \theta_2)$ is to be estimated based on $X = (X_1, X_2)$. Here $\Theta_o = \Omega_o = R^2$, $\Theta = \Omega = \{\theta \mid \theta_1 \leq \theta_2\}$ and estimators $\delta(X) = (\delta_1(X), \delta_2(X))$ satisfy $P_\theta(\delta_1(X) \leq \delta_2(X)) = 1$ for all $\theta \in \Theta$. As another example, let $X_i \sim^{ind} \mathcal{N}(\theta, \lambda_i)$, $i = 1, \ldots, k$ with all parameters unknown and $0 < \lambda_1 \leq \ldots \leq \lambda_k$. Then θ is to be estimated based on $X = (X_1, \ldots, X_k)$. Here $K = k+1$, $M = 1$, $\Omega_o = \{(\theta, \lambda_1, \ldots, \lambda_k) \mid -\infty < \theta < \infty, \lambda_i > 0, i = 1, \ldots, k\}$, $\Omega = \{(\theta, \lambda_1, \ldots, \lambda_k) \mid -\infty < \theta < \infty, 0 < \lambda_1 \leq \ldots \leq \lambda_k\}$, and $\Theta_o = \Theta = R^1$.

Not every author on the subject of estimation in restricted parameter spaces restricts his estimators of θ to those satisfying (2.3). Some authors ask, for some or all of their estimators, only that they satisfy

$$P_{\theta,\lambda}(\delta(X) \in \Theta_o) = 1 \text{ for all } (\theta, \lambda) \in \Omega. \tag{2.5}$$

Others do not say what they consider to be an estimator, but their definition can sometimes be obtained from the properties they prove their estimators to have. A summary of opinions on whether estimators should satisfy (2.3) can be found in Blyth (1993). Here I only quote Hoeffding (1983) on the subject of restricting estimators of θ to those in $\mathcal{D}$. He calls such estimators "range-preserving" and says "The property of being range-preserving is an essential property of an estimator, a sine qua non. Other properties, such as unbiasedness, may be desirable in some situations, but an unbiased estimator that is not range-preserving should be ruled out as an estimator." – a statement with which I agree.

Let $L(d, \theta)$ be the loss incurred when $d \in \Theta$ is used to estimate θ and $\theta \in \Theta$ is the true value of the parameter to be estimated. It is assumed that L is of the form

$$L(d, \theta) = \sum_{i=1}^{M} L_i(d_i, \theta_i), \tag{2.6}$$

where, for each $i = 1, \ldots, M$, all $y = (y_1, \ldots, y_M) \in \Theta$ and all $(\theta_1, \ldots, \theta_M) \in \Theta$,

$$\left.\begin{array}{l} i) \ \ L_i(y_i, \theta_i) \text{ is bowl-shaped in } \theta_i, \\[4pt] ii) \ \ L_i(y_i, \theta_i) \geq 0 \text{ and } L_i(\theta_i, \theta_i) = 0, \\[4pt] iii) \ \ L_i(y_i, \theta_i) \text{ is convex in } y_i. \end{array}\right\} \tag{2.7}$$

These properties of the loss function, together with the convexity of Θ, imply that the class of non-randomized estimators is essentially complete in the class of all estimators with respect to Ω in the sense that, for every randomized estimator δ, there exists a non-randomized one δ' with $\mathcal{E}_{\theta,\lambda} L(\delta'(X), \theta) \leq L(\delta(X), \theta)$ for all $(\theta, \lambda) \in \Omega$. So one can restrict oneself to non-randomized estimators.

Examples of loss functions of the form (2.6) with the properties (2.7) are

1) the class of weighted p^{th}-power loss functions where

$$L(d, \theta) = \sum_{i=1}^{M} |d_i - \theta_i|^p w_i(\theta).$$

Here $p \geq 1$ and the $w_i(\theta)$ are known functions of θ which are, for each $i = 1, \dots, M$, strictly positive on Θ. Special cases of this loss function are (i) squared-error loss with $p = 2$ and $w_i(\theta) = 1$ for all $i = 1, \dots, M$ and (ii) scale-invariant squared-error loss with $p = 2$ and $w_i(\theta) = 1/\theta_i^2$, which can be used when $\theta_i > 0$ for all $\theta \in \Theta$, as is, e.g., the case in scale-parameter estimation problems;

2) the class of linex loss functions where

$$L(d, \theta) = \sum_{i=1}^{M} \left(e^{w_i(\theta)(d_i - \theta_i)} - w_i(\theta)(d_i - \theta_i) - 1 \right).$$

Here the $w_i(\theta)$ are known functions of θ with, for each $i = 1, \dots, M$, $w_i(\theta) \neq 0$ for all $\theta \in \Theta$.

In problems with $M \geq 2$ quadratic loss is sometimes used. It generalizes squared-error loss and is given by

$$L(d, \theta) = (d - \theta)' A(d - \theta), \tag{2.8}$$

where A is a known $M \times M$ positive definite matrix. For instance, when $X \sim \mathcal{N}_M(\theta, \Sigma)$, Σ known and positive definite, with the vector θ to be estimated, taking $A = \Sigma^{-1}$ is equivalent to estimating the vector $\Sigma^{-1/2}\theta$ with squared-error loss based on $Y = \Sigma^{-1/2} X$.

The risk function of an estimator δ of θ is, for $(\theta, \lambda) \in \Omega$, given by $R(\delta, (\theta, \lambda)) = \mathcal{E}_{\theta,\lambda} L(\delta(X), \theta)$ and estimators are compared by comparing their risk functions. An estimator δ is called inadmissible in a class $\mathcal{C}$ of estimators for estimating θ if there exists an estimator $\delta' \in \mathcal{C}$ dominating it on Ω, i.e., if there exists an estimator $\delta' \in \mathcal{C}$ with

$$R(\delta', (\theta, \lambda)) \leq R(\delta, (\theta, \lambda)) \text{ for all } (\theta, \lambda) \in \Omega \text{ and}$$

$$R(\delta', (\theta, \lambda)) < R(\delta, (\theta, \lambda)) \text{ for some } (\theta, \lambda) \in \Omega$$

and an estimator δ is admissible when it is not inadmissible. Further, an estimator δ of θ is called minimax in a class $\mathcal{C}$ of estimators of θ if it minimizes, among estimators $\delta' \in \mathcal{C}$, $\sup_{(\theta,\lambda)\in\Omega} R(\delta', (\theta, \lambda))$.

In the literature on estimation in restricted parameter spaces two definitions of admissibility and minimaxity are used. In each of the definitions the risk functions of estimators of θ are compared on Ω. However, in one definition the estimators under consideration (i.e., the above class $\mathcal{C}$) are those in $\mathcal{D}$, while in the other definition the estimators are those in $\mathcal{D}_o = \{\delta \mid (2.5) \text{ is satisfied}\}$. Or, to say this another way, by the first definition an estimator $\delta \ (\in \mathcal{D})$ is inadmissible when there exists an estimator $\delta' \in \mathcal{D}$ which dominates δ on Ω. And an estimator $\delta \ (\in \mathcal{D})$ is minimax when it minimizes, among estimators in $\mathcal{D}$, $\sup\{R(\delta, (\theta, \lambda)) \mid (\theta, \lambda) \in \Omega\}$. By the second definition, an estimator δ $(\in \mathcal{D}_o)$ is inadmissible if there exists an estimator $\delta' \in \mathcal{D}_o$ which dominates it on Ω. And an estimator $\delta \ (\in \mathcal{D}_o)$ is minimax if it minimizes, among the estimators in $\mathcal{D}_o$, $\sup\{R(\delta, (\theta, \lambda)) \mid (\theta, \lambda) \in \Omega\}$. It is hereby assumed, for the second pair of definitions, that the loss function (2.6)

$$\left.\begin{array}{l} i) \text{ is defined for } \theta \in \Theta \text{ and } d \in \Theta_o \\[2mm] ii) \text{ satisfies (2.7) with } \theta \in \Theta \text{ and } (y_1, \ldots, y_M) \in \Theta_o. \end{array}\right\} \quad (2.9)$$

These two notions of admissibility and minimaxity will be called, respectively, $(\mathcal{D}, \Omega)$- and $(\mathcal{D}_o, \Omega)$-admissibility and minimaxity and the corresponding estimation problems will be called, respectively, the $(\mathcal{D}, \Omega)$- and the $(\mathcal{D}_o, \Omega)$-problems. In this monograph estimators satisfy (unless specifically stated otherwise) (2.3) and admissibility and minimaxity mean $(\mathcal{D}, \Omega)$-admissibility and minimaxity.

The following relationships exist between $(\mathcal{D}, \Omega)$- and $(\mathcal{D}_o, \Omega)$-admissibility and minimaxity:

$$\delta \in \mathcal{D}, \delta \text{ is } (\mathcal{D}_o, \Omega)\text{-admissible} \implies \delta \text{ is } (\mathcal{D}, \Omega)\text{-admissible.} \quad (2.10)$$

Further,

$$\delta \in \mathcal{D}, \delta \text{ is } (\mathcal{D}_o, \Omega)\text{-minimax} \implies \left\{\begin{array}{l} \delta \text{ is } (\mathcal{D}, \Omega)\text{-minimax,} \\[2mm] M(\mathcal{D}, \Omega) = M(\mathcal{D}_o, \Omega), \end{array}\right. \quad (2.11)$$

where $M(\mathcal{D}, \Omega)$ and $M(\mathcal{D}_o, \Omega)$ are the minimax values for the classes $\mathcal{D}$ and $\mathcal{D}_o$ and the parameter space Ω.

Now note that, for weighted squared-error loss, the class of estimators $\mathcal{D}$ is essentially complete in $\mathcal{D}_o$ with respect to the parameter space Ω in the sense that, for every $\delta \in \mathcal{D}_o$, $\delta \notin \mathcal{D}$ there exists a $\delta' \in \mathcal{D}$ dominating it on Ω. This dominator is obtained by minimizing, for each $x \in \mathcal{X}$, $L(\delta(x), \theta)$ in θ for

$\theta \in \Theta$. This essential completeness also holds for a rectangular Θ when the loss function is given by (2.6) and satisfies (2.9). Further, one can dominate δ by using what Stahlecker, Knautz and Trenkler (1996) call the "minimax adjustment technique". Their dominator – δ', say – is obtained by minimizing, for each $x \in \mathcal{X}$,

$$H(d) = \sup_{\theta \in \Theta}(L(d,\theta) - L(\delta(x),\theta))$$

for $d \in \Theta$. When $\delta(x) \in \Theta$, $H(d) \geq 0$ because

i) $H(\delta(x)) = 0$, so $\inf_{d \in \Theta} H(d) \leq 0$;

ii) $\inf_{d \in \Theta} H(d) < 0$ contradicts the fact that, for each $d \in \Theta$,

$$H(d) = \sup_{\theta \in \Theta}(L(d,\theta) - L(\delta(x),\theta))$$

$$\geq L(d, \delta(x)) - L(\delta(x), \delta(x)) = L(d, \delta(x)) \geq 0.$$

So, when $\delta(x) \in \Theta$, $d = \delta(x)$ is a minimizer of $H(d)$. When $\delta(x)$ is not in Θ, assume that a minimizer exists. Then we have, for all $\theta \in \Theta$ and all $x \in \mathcal{X}$,

$$\left.\begin{array}{l} L(\delta'(x),\theta) - L(\delta(x),\theta) = \\[2mm] \inf_{d \in \Theta} \sup_{\theta \in \Theta}(L(d,\theta) - L(\delta(x),\theta)) = \\[2mm] \sup_{\theta \in \Theta} \inf_{d \in \Theta}(L(d,\theta) - L(\delta(x),\theta)) = \\[2mm] \sup_{\theta \in \Theta}(-L(\delta(x),\theta)) = -\inf_{\theta \in \Theta} L(\delta(x),\theta) \leq 0, \end{array}\right\} \quad (2.12)$$

where it is assumed that inf and sup can be interchanged.

Essential completeness of $\mathcal{D}$ in $\mathcal{D}_o$ with respect to Ω, together with (2.10) gives

$$\delta \text{ is } (\mathcal{D}, \Omega)\text{-admissible} \iff \delta \in \mathcal{D}, \delta \text{ is } (\mathcal{D}_o, \Omega)\text{-admissible}. \quad (2.13)$$

Further, using (2.11) and the essential completeness of $\mathcal{D}$ in $\mathcal{D}_o$ with respect to Ω, one obtains

$$\left.\begin{array}{l} \delta \text{ is } (\mathcal{D}, \Omega)\text{-minimax} \iff \delta \in \mathcal{D}, \delta \text{ is } (\mathcal{D}_o, \Omega)\text{-minimax.} \\[2mm] M(\mathcal{D}, \Omega) = M(\mathcal{D}_o, \Omega). \end{array}\right\} \quad (2.14)$$

From (2.13) and (2.14) it is seen that studying the $(\mathcal{D}_o, \Omega)$-problem can be very helpful for finding admissibility and minimaxity results for the $(\mathcal{D}, \Omega)$-problem.

Another problem whose admissibility and minimaxity results can be helpful for our problem is the "unrestricted problem" where estimators of θ are restricted to Θ_o and compared on Ω_o. Then we have (still assuming that (2.2) holds)

$$\delta \in \mathcal{D}_o \implies P_{\theta,\lambda}(\delta(X) \in \Theta_o) = 1 \text{ for all } (\theta, \lambda) \in \Omega_o,$$

so that this estimation problem can, and will, be called the $(\mathcal{D}_o, \Omega_o)$-problem. Obviously,

$$\delta \in \mathcal{D}_o, \delta \text{ is } (\mathcal{D}_o, \Omega)\text{-admisible} \implies \delta \in \mathcal{D}_o, \delta \text{ is } (\mathcal{D}_o, \Omega_o)\text{-admissible.} \quad (2.15)$$

Also, because $\Omega \subset \Omega_o$,

$$M(\mathcal{D}_o, \Omega) \leq M(\mathcal{D}_o, \Omega_o) \quad (2.16)$$

which, together with (2.14), gives

$$M(\mathcal{D}, \Omega) = M(\mathcal{D}_o, \Omega) \leq M(\mathcal{D}_o, \Omega_o). \quad (2.17)$$

One can now ask the question: when does

$$M(\mathcal{D}_o, \Omega) = M(\mathcal{D}_o, \Omega_o) \quad (2.18)$$

or, equivalently,

$$M(\mathcal{D}, \Omega) = M(\mathcal{D}_o, \Omega_o) \quad (2.19)$$

or, equivalently

$$M(\mathcal{D}, \Omega) = M(\mathcal{D}_o, \Omega) = M(\mathcal{D}_o, \Omega_o) \quad (2.20)$$

hold? Or – are there cases where restricting the parameter space does not reduce the minimax value of the problem?

Examples where (2.2) and (2.20) hold can be found in Chapter 4, sections 4.2, 4.3 and 4.4. Those in the sections 4.3 and 4.4 are examples where either all the θ_i are lower-bounded or $\Theta = \{\theta \mid \theta_1 \leq \ldots, \leq \theta_k\}$. In the example in Section 4.2, $X \sim \text{Bin}(n, \theta)$ with $\theta \in [m, 1-m]$ for a small known $m \in (0, 1/2)$.

An example where (2.2) is satisfied but (2.20) does not hold is the estimation of a bounded normal mean with $\Theta_o = (-\infty, \infty)$, squared-error loss and known variance. When $\Theta = [-m, m]$ for some positive known m, Casella and Strawderman (1981) give the values of $M(\mathcal{D}, \Omega)$ for several values of m. For example, for $m = .1, .5, 1$ and a normal distribution with variance 1, the minimax risks are, respectively $.010, .199, .450$, while, of course, $M(\mathcal{D}_o, \Omega_o) = 1$, showing that restricting the parameter space to a compact set can give very substantial reductions in the minimax value of the problem. These results are discussed in Chapter 4, Section 4.2 together with other cases where the three minimax values are not equal.

As already noted, the above relationships between admissibility and minimaxity results for the $(\mathcal{D}, \Omega)$- and $(\mathcal{D}_o, \Omega)$-problems show that solving a $(\mathcal{D}_o, \Omega)$-problem can be very helpful toward finding a solution to the corresponding $(\mathcal{D}, \Omega)$-problem. But authors who publish results on a $(\mathcal{D}_o, \Omega)$-problem are not always clear about why they do so. Is it as a help for solving the corresponding $(\mathcal{D}, \Omega)$-problem, or do they consider statistics not satisfying (2.3) to be estimators and are not really interested in the corresponding $(\mathcal{D}, \Omega)$-problem? In this monograph some papers are included which look at $(\mathcal{D}_o, \Omega)$-problems. Their results are clearly identified as such, but their relationship to the corresponding $(\mathcal{D}, \Omega)$-problems is not always commented on.

Remark 2.1. Note that, when $\delta \in \mathcal{D}_o$ is $(\mathcal{D}_o, \Omega_o)$-minimax and $\delta' \in \mathcal{D}$ dominates δ on Ω, one cannot conclude that δ' is $(\mathcal{D}, \Omega)$-minimax. But Dykstra (1990) seems, in his Example 3, to draw this conclusion.

Remark 2.2. The above definition of the minimax adjustment technique is not the one used by Stahlecker, Knautz and Trenkler (1996). They minimize, for each $x \in \mathcal{X}$, $H(d)$ for $d \in R^k$. Such a minimizer is not necessarily constrained to Θ. But assuming they meant to minimze over Θ, their reasoning is incorrect.

In most papers on restricted-parameter-space estimation the models considered satisfy (2.2), but several models where this condition is not satisfied are rather extensively studied. Three examples are the k-sample problems where $X_{i,1}, \ldots, X_{i,n_i}, i = 1, \ldots, k$, are independent and $X_{i,j}, j = 1, \ldots, n_i$ have either a $\mathcal{U}(0, \theta_i)$ distribution or a $\mathcal{U}(\theta_i - 1, \theta_i + 1)$ distribution or an exponential distribution on (θ_i, ∞). Suppose, in the first uniform case, that $\theta = (\theta_1, \ldots, \theta_k)$ is to be estimated when $\theta \in \Theta$, where Θ is a closed convex subset of R_+^k with a non-empty interior. Then $M = k = K$, $\Omega_o = \Theta_o = R_+^k$ and $\Omega = \Theta$. Given that we know for sure, i.e., with $P_\theta = 1$ for all $\theta \in R_+^k$, that $Y_i = \max_{1 \leq j \leq n_i} X_{i,j} \leq \theta_i, i = 1, \ldots, k$, estimators δ of θ "should", in addition to being restricted to Θ, satisfy the "extra" restriction that $\delta_i(Y) \geq Y_i$, $i = 1, \ldots, k$, where $Y = (Y_1, \ldots, Y_k)$. To say it more precisely, δ should satisfy

$$P_\theta(\delta(Y) \in \Theta_Y) = 1 \text{ for all } \theta \in \Theta, \tag{2.21}$$

where

$$\Theta_Y = \{\theta \in \Theta \mid \theta_i \geq Y_i, i = 1, \ldots, k\}.$$

Let $\mathcal{D}'$ be the class of estimators satisfying (2.21) and let the $(\mathcal{D}', \Theta)$-problem be the problem where estimators are in $\mathcal{D}'$ and are compared on Θ. The unrestricted problem in this case is the problem where estimators satisfy

$$P_\theta(\delta(Y) \in \Theta_{o,Y}) = 1 \text{ for all } \theta \in R_+^k, \tag{2.22}$$

where

$$\Theta_{o,Y} = \{\theta \in R_+^k \mid \theta_i \geq Y_i, i = 1, \ldots, k\}$$

and estimators are compared on R_+^k. Call this problem the $(\mathcal{D}'_o, R_+^k)$-problem. And then there is the problem studied by those who do not insist that their estimators are restricted to Θ, i.e., the $(\mathcal{D}'_o, \Theta)$-problem where estimators satisfy (2.22) and are compared on Θ.

From the above definitions it follows that $\Theta_Y \subset \Theta_{o,Y}$ with $P_\theta = 1$ for all $\theta \in \Theta$ and that Θ_Y and $\Theta_{o,Y}$ are both closed and convex with a non-empty interior. So, $\mathcal{D}' \subset \mathcal{D}'_o$ and, under the same conditions on the loss function as before, $\mathcal{D}'$ is essentially complete in $\mathcal{D}'_o$ with respect to Θ. This shows that (2.13) and (2.14) hold with $\mathcal{D}$ (resp. $\mathcal{D}_o$) replaced by $\mathcal{D}'$ (resp. $\mathcal{D}'_o$). With these same replacements, (2.15) holds for the $(\mathcal{D}'_o, R_+^k)$- and the $(\mathcal{D}'_o, \Theta)$-problems so that (see (2.17)),

$$M(\mathcal{D}', \Theta) = M(\mathcal{D}'_o, \Theta) \leq M(\mathcal{D}'_o, R_+^k). \tag{2.23}$$

An example where these three minimax values are equal is given in Chapter 4, Section 4.3.

Similar remarks and results hold for the other uniform case and for the exponential case, as well as for cases with nuisance parameters.

In order to simplify the notation, $(\mathcal{D}, \Omega)$, $(\mathcal{D}_o, \Omega)$ and $(\mathcal{D}_o, \Omega_o)$ are used for the three problems (with the Ω's replaced by Θ's in case there are no nuissance parameters), whether (2.2) is satisfied or not: i.e., the primes are left off for cases like the uniform and exponentail ones above. And "δ satisfies (2.3)" stands for "δ satisfies (2.3) or (2.21)", as the case may be.

Quite a number of papers on such uniform and exponential models are discussed in this monograph. In most of them the "extra" restriction is taken into account but, as will be seen in Chapter 5, Sections 5.2 and 5.3, in two cases authors propose and study estimators which do not satisfy it.

Two more remarks about admissibility and minimaxity for the three problems: (i) if δ is $(\mathcal{D}, \Omega)$-inadmissible as well as $(\mathcal{D}, \Omega)$-minimax, then every $\delta' \in \mathcal{D}$ which dominates δ on Ω is also $(\mathcal{D}, \Omega)$-minimax. This also holds with $(\mathcal{D}, \Omega)$ replaced by $(\mathcal{D}_o, \Omega)$ as well with $(\mathcal{D}, \Omega)$ replaced by $(\mathcal{D}_o, \Omega_o)$; (ii) if δ is $(\mathcal{D}_o, \Omega)$-minimax and $\mathcal{D}$ is essentially complete in $\mathcal{D}_o$ with respect to Ω then there exists a $\delta' \in \mathcal{D}$ which is $(\mathcal{D}, \Omega)$-minimax, because the essential completeness implies that $M(\mathcal{D}, \Omega) = M(\mathcal{D}_o, \Omega)$.

Universal domination is another criterion for comparing estimators. It was introduced by Hwang (1985) for the case where $\Theta = \Omega = R^k$ and by this criterion an estimator δ' universally dominates an estimator δ on the parameter space Θ with respect to a class $\mathcal{C}$ of loss functions L if

$$\mathcal{E}_\theta L(\delta'(X), \theta) \leq \mathcal{E}_\theta L(\delta(X), \theta) \quad \text{for all } \theta \in \Theta \text{ and all } L \in \mathcal{C}$$

and, for a particular loss function $\in \mathcal{C}$, the risk functions are not identical. An estimator δ is called universally admissible if no such δ' exists.

Hwang (1985) takes the class $\mathcal{C}$ to be the class of all nondecreasing functions of the generalized Euclidean distance $|d - \theta|_D = ((d - \theta)'D(d - \theta))^{1/2}$ where D is a given non-negative definite matrix. This implies, as he shows, that δ' universally dominates δ if and only if δ' stochastically dominates δ, i.e. if and only if

$$P_\theta(|\delta'(X) - \theta|_D \geq c) \leq P_\theta(|\delta(X) - \theta|_D \geq c) \quad \text{for all } c > 0 \text{ and all } \theta \in \Theta$$

$$P_\theta(|\delta'(X) - \theta|_D \geq c) < P_\theta(|\delta(X) - \theta|_D \geq c) \quad \text{for some } (c, \theta), c > 0, \theta \in \Theta.$$

He further shows that, if δ is admissible with respect to a particular loss function L_o which is a strictly increasing function of $|d - \theta|_D$ and the risk function of δ for this loss function is finite for all $\theta \in \Theta$, then δ is universally admissible for this D and Θ. Equivalently, if δ is universally inadmissible with respect to a D and Θ, then δ is inadmissible under any strictly increasing loss $L_o(|d - \theta|_D)$ with a risk which is finite for all $\theta \in \Theta$.

Still another criterion for comparing estimators is Pitman closeness (also called Pitman nearness). For two estimators δ_1 and δ_2 of $\theta \in \Theta \subset R^1$, Pitman (1937) defines, for cases where $K = M$, their closeness by

$$P_\theta(|\delta_1(X) - \theta| < |\delta_2(X) - \theta|) \quad \theta \in \Theta. \tag{2.24}$$

Then, assuming that $P_\theta(|\delta_1(X) - \theta| = |\delta_2(X) - \theta|) = 0$ for all $\theta \in \Theta$, δ_1 is Pitman-closer to θ than δ_2 when (2.24) is $\geq 1/2$. Pitman (1937) notes that Pitman closeness comparisons are not necessarily transitive. For three estimators δ_i, $i = 1, 2, 3$, one can have δ_1 Pitman-closer to θ than δ_2, δ_2 Pitman-closer to θ than δ_3 and δ_3 Pitman-closer to θ than δ_1. It is also well-known that Pitman-closeness comparisons do not necessarily agree with risk-function comparisons. One can have, e.g., δ_1 Pitman-closer to θ than δ_2 while δ_2 dominates δ_1 for squared-error loss. In Chapter 5, Section 5.3, several Pitman-closeness comparisons in restricted parameter spaces are presented and compared with risk-function comparisons. Much more on Pitman closeness, in particular on its generalization to $k \geq 2$, can be found in Keating, Mason and Sen (1993).

One of the various estimators discussed in this monograph is the so-called Pitman estimator. The name comes from Pitman (1939). He proposes and studies Bayes estimators with respect to a uniform prior. His parameters are either location parameters $\theta \in \Theta_o = (-\infty, \infty)$, for which he uses squared-error loss and a uniform prior on Θ, or scale parameters $\theta \in \Theta_o = (0, \infty)$, for which he uses scale-invariant-squared-error loss and a uniform prior for $\log \theta$ on $(-\infty, \infty)$. But the name "Pitman estimator" is now used by many authors, and is used in this monograph, for any Bayes estimator with respect to a uniform prior for θ or for a function $h(\theta)$ for $\theta \in \Theta$ or Θ_o. Some of the properties of the original Pitman estimators are summarized in Chapter 4, Section 4.1.

In many restricted-parameter-space estimation problems considered in the literature the problem does not contain any nuisance parameters, the problem is a k-sample problem with independent samples from distributions $F_i(x, \theta_i)$, $i = 1, \ldots, k$ and Θ $(= \Omega)$ is determined by inequalities among the components θ_i of θ. The most common examples are the simple-order restriction where $\Theta = \{\theta \mid \theta_1 \leq \ldots \leq \theta_k\}$, the simple-tree-order restriction with $\Theta = \{\theta \mid \theta_1 \leq \theta_i, i = 2, \ldots, k\}$, the umbrella-order restriction with $\Theta = \{\theta \mid$ for some $i_o, 1 < i_o < k, \theta_i \leq \theta_{i_o}$ for all $i \neq i_o\}$ and the loop-order restriction with, for $k = 4$ e.g., $\Theta = \{\theta \mid \theta_1 \leq \theta_2 \leq \theta_4, \theta_1 \leq \theta_3 \leq \theta_4\}$. The simple-tree-order restriction is a special case of the rooted-tree-order restriction where each θ_i, except one of them (θ_1, say, the root), has exactly one immediate predecessor and the root has none. Here, θ_j is an immediate predecessor of θ_i $(i \neq j)$ when $\theta \in \Theta$ implies $\theta_j \leq \theta_i$ but there does not exist an $l, l \neq i, l \neq j$, with $\theta_j \leq \theta_l \leq \theta_i$. So, the simple-tree order is a tree order where all θ_i have the root as their unique immediate predecessor. Another Θ for which results have been obtained is the upper-star-shaped restriction, also called the increasing-in-weighted-average restriction, where

$$\Theta = \{\theta \mid \bar{\theta}_1 \leq \bar{\theta}_2 \leq \ldots \leq \bar{\theta}_k\},$$

with $\bar{\theta}_i = \sum_{j=1}^{i} w_j \theta_j / \sum_{j=1}^{i} w_j$ for given positive weights w_i. Note that this Θ is equivalent to

$$\Theta = \{\theta \mid \bar{\theta}_i \leq \theta_{i+1}, i = 1, \ldots, k - 1\}.$$

Finally, when $\theta_1, \ldots, \theta_k$ are order-restricted, θ_i is a node when, for each $j \neq i$, $\theta_j \leq \theta_i$ or $\theta_j \geq \theta_i$. For instance, when $k = 5$ and

$$\Theta = \{\theta \mid \theta_j \leq \theta_3, j = 1, 2, \theta_l \geq \theta_3, l = 4, 5\},$$

θ_3 is the only node and when

$$\Theta = \{\theta \mid \theta_j \leq \theta_3, j = 1, 2, \theta_3 \leq \theta_4 \leq \theta_5\}$$

then θ_3, θ_4 and θ_5 are nodes, but θ_1 and θ_2 are not. Nodes are important for some estimation problems presented in Chapter 5.

3

(In)admissibility and dominators

In this chapter results are presented on (in)admissibility of estimators of θ (satisfying (2.3)) for the case where the problem does not contain any nuisance parameters. So, in the notation of Chapter 2, $M = K$, $\Omega_o = \Theta_o$ and $\Omega = \Theta$ and the notation M for the number of parameters to be estimated is changed to k. Some of these results give sufficient conditions for inadmissibility of estimators of θ for a particular family of distributions. For example, when is an estimator of the expectation parameter in a one-dimensional exponential-family distribution inadmissible, when Θ is a closed, convex subset of the (open) natural parameter space and the loss is squared error? Other results are for particular estimators: e.g., is the MLE of θ, when X has a logistic distribution with mean θ, inadmissible for squared error loss when $\Theta = \{\theta \mid m_1 \leq \theta \leq m_2\}$ for known m_1 and m_2 with $-\infty < m_1 < m_2 < \infty$? Dominators for some of the inadmissible estimators are given.

3.1 Results for the exponential family

We start with what is possibly the earliest result on inadmissiblity in restricted parameter spaces. It can be found in Hodges and Lehmann (1950). Let $X \sim \text{Bin}(1, \theta)$ with $1/3 \leq \theta \leq 2/3$ and let the loss be squared error. Then Hodges and Lehmann show that the MLE of θ is inadmissible and dominated by every estimator δ satisfying $1/3 \leq \delta(0) \leq \delta(1) = 1 - \delta(0) \leq 2/3$. Another early result can be found in Sacks (1960, 1963 (p. 767)). Let $X \sim \mathcal{N}(\theta, 1)$, where we know that $\theta \geq 0$ and let the loss for estimating θ be squared error. Then the MLE $(= \max(0, X))$ is inadmissible. This result is a special case of a much more general result proved by Sacks . The same kind of result can be found in Brown (1986, Theorem 4.23) and we will state it in the form Brown gives it. The setting is a one-parameter exponential family with density

$$p_\xi(x) = C(\xi) \exp(x\,\xi), \tag{3.1}$$

with respect to a σ-finite measure ν. The natural parameter space N is open, the loss is squared error and the expectation parameter $\theta = \eta(\xi) = \mathcal{E}_\xi X$ is to be estimated. Brown shows that, when ξ is restricted to Ξ, a closed convex subset of the natural parameter space, then an admissible estimator δ of θ is non-decreasing. Further, if

$$\left. \begin{aligned} I_\delta = \\ \{x \mid \nu(\{y \mid y > x, \delta(y) \in S^o\}) > 0, \nu(\{y \mid y < x, \delta(y) \in S^o\}) > 0\}, \end{aligned} \right\} \quad (3.2)$$

where S^o is the interior of S, the closed convex support of ν, and δ is admissible, then there exists a finite measure V on Ξ such that, for all $x \in I_\delta$,

$$\delta(x) = \frac{\displaystyle\int \frac{\eta(\xi)}{1 + |\eta(\xi)|} e^{x\,\xi} \, dV(\xi)}{\displaystyle\int \frac{1}{1 + |\eta(\xi)|} e^{x\,\xi} \, dV(\xi)}. \quad (3.3)$$

Obviously, Sack's (1963) result for the lower-bounded normal mean follows from this result of Brown.

There are many other cases where Brown's Theorem 4.23 can be used to prove inadmissibility of estimators for restricted-parameter-space problems. Some examples are:

i) $X \sim \mathcal{N}(\theta, 1)$ with $\theta = \xi$, $N = \{\xi \mid -\infty < \xi < \infty\}$ and $\Theta = \{\theta \mid m_1 \le \theta \le m_2\}$, $-\infty < m_1 < m_2 < \infty$;

ii) $X \sim \Gamma(\alpha, \theta)$ for known $\alpha > 0$, $\xi = \theta^{-1}$, $\alpha\theta = \mathcal{E}_{\alpha,\theta}X$, $N = \{\xi \mid \xi > 0\}$ and $\Theta = \{\theta \mid m_1 \le \theta \le m_2\}$ with $0 < m_1 < m_2 \le \infty$;

iii) $X \sim \text{Bin}(n, \theta)$ with $\xi = \log(\theta/(1-\theta))$, $N = \{\xi \mid -\infty < \xi < \infty\}$ and $\Theta = \{\theta \mid m_1 \le \theta \le m_2\}$ for $0 < m_1 < m_2 < 1$.

In each of these cases m_1 and m_2 are known, the loss is squared error and the expectation parameter is to be estimated. Then the MLE is an example of an estimator which is "not smooth enough" to satisfy Brown's necessary condition for admissibilty, but there are of course many others.

Further, any of these results can be extended to the case where $X_1, \ldots, X_k$ are independent and, for $i = 1, \ldots, k$, X_i has density (3.1) with ξ replaced by ξ_i. Restricting $(\xi_1, \ldots, \xi_k)$ to a closed rectangular subset of N^k and using squared-error loss, the inadmissibility of, e.g., the MLE of the vector of the expectation parameters then follows from Brown's Theorem 4.23. This theorem of Brown cannot be applied when Θ is not a rectangle, but for the case where, e.g., $X \sim \mathcal{N}_k(\xi, I)$ with ξ restricted to a closed, convex subset of R^k and squared-error loss, inadmissibility of the MLE of the vector ξ follows from Theorem 4.16 of Brown (1986). In this theorem Brown gives necessary conditions for the admissibility of an estimator of the canonical parameter of a k-parameter exponential family with density

$$p_\xi(x_1, \ldots, x_k) = C_1(\xi) \exp\left(\sum_{i=1}^{k} x_i\, \xi_i\right), \qquad (3.4)$$

where $\xi = (\xi_1, \ldots, \xi_k)$. He shows that δ is an admissible estimator of ξ when ξ is restricted to Ξ, a closed convex subset of $\bar{N}$, only if there exits a measure H on Ξ such that

$$\delta(x) = \frac{\int \xi\, e^{x\,\xi} dH(\xi)}{\int e^{x\,\xi} dH(\xi)} \qquad \text{for } x \in S^o \text{ a.e. } \nu, \qquad (3.5)$$

where $x = (x_1, \ldots, x_k)$. An example of this normal-mean result is the case where $X_1, \ldots, X_k$ are independent with $X_i \sim N(\xi_i, 1)$, $i = 1, \ldots, k$ and $\xi_1 \leq \ldots \leq \xi_k$.

There are cases within the exponential family where Brown's Theorem 4.23 can not be used, but where (in)admissibility results have been obtained. For instance:

1) The case where $X_1, \ldots, X_k$, $k \geq 2$, are independent with, for $i = 1, \ldots, k$, $X_i \sim \text{Bin}(n_i, \theta_i)$ and $\Theta = \{\theta \mid \theta_1 \leq \ldots \leq \theta_k\}$, where $\theta = (\theta_1, \ldots, \theta_k)$. Sackrowitz and Strawderman (1974) show that, for squared-error loss, the MLE of θ is admissible if and only if

$$\sum_{i=1}^{k} n_i < 7 \text{ or } (k = 2, n_i = 1 \text{ for some } i \in \{1, 2\}) \text{ or } (k = 3, n_1 = n_3 = 1).$$

They also have results for weighted squared-error loss;

2) As already mentioned in the Introduction, Charras and van Eeden (1991a) show that, for $X \sim \text{Bin}(n, \theta)$ with $\theta \in [0, m]$ for some known $m \in (0, 1)$ and squared-error loss, the MLE of θ is admissible when $n \geq 2$ and $0 < nm \leq 2$. Funo (1991, Theorem 3.1) shows inadmissibility of the MLE when $nm > 2$. Funo also has results for the estimation of restricted multinomial parameters. He bases his proofs on a complete class theorem of Brown (1981) which is concerned with inference problems when the sample space is finite and the parameter space might be restricted;

3) For the case where X has a Poisson distribution with mean $\theta \in [0, m]$, $0 < m < \infty$, Charras and van Eeden (1991a) show, among other things, that an estimator δ with

$$\delta(0) = 0 < \delta(1) \leq m \text{ and } \delta(x) = m \text{ for } x \geq 2$$

is admissible for estimating θ with squared error loss;

4) Shao and Strawderman (1994) consider the estimation, for squared-error loss, of the mean of a power-series distribution under restrictions on the parameter space. They give admissibility results for the MLE and note that some of the above-given results of Charras and van Eeden (1991a) for the binomial and Poisson cases overlap with theirs.

In the first example above, Brown's Theorem 4.23 does not apply because $k \geq 2$ and Θ is not a rectangle. In each of the examples in 2) - 4) this theorem of Brown does not apply because Θ is not a closed convex subset of the (open) natural parameter space.

Remark 3.1. The phenomenon seen in 1) and 2) above that an estimator is admissible for small sample sizes and inadmissible for larger ones also occurs in unrestricted parameter spaces. Brown, Chow and Fong (1992), for example, show that, for $X \sim Bin(n, \theta)$, the MLE of $\theta(1 - \theta)$ is, for squared-error loss, admissible for $n \leq 5$ and inadmissible for $n \geq 6$.

Remark 3.2. My above statements of Brown's theorems 4.23 and 4.16 are not identical to Brown's in that I added the condition that the parameter space is a closed, convex subset of N for Theorem 4.23 and of $\bar{N}$ for Theorem 4.16 – conditions Brown uses in his proofs, but apparently forgot to mention in the statements of his theorems.
Further, Brown's proofs – in particular his proof of his Theorem 4.23 – are not complete and need more precision in some places. But, at the moment, I am convinced that they can be fixed and that these theorems as I state them are correct.

3.2 Results of Moors

Moors (1981, 1985) considers the general problem as described at the beginning of Chapter 2. He estimates $h(\theta)$ for a given h taking values in R^k. His results are presented here for the special case where $h(\theta) = \theta$ for all $\theta \in \Theta_o$.

Moors uses squared-error loss and assumes that the problem of estimating θ is invariant with respect to a finite group of transformations from $\mathcal{X}$ to $\mathcal{X}$. He also assumes that the induced group $\tilde{G}$ satisfies

$$\tilde{g}(\alpha d_1 + \beta d_2) = \alpha \tilde{g}(d_1) + \beta \tilde{g}(d_2) \quad \text{for all } \alpha, \beta \in R, \text{ all } d_1, d_2 \in \Theta, \text{ all } \tilde{g} \in \tilde{G},$$

that all $g \in G$ are measure-preserving (i.e., $\nu(g^{-1}(A)) = \nu(A)$ for all $A \in \mathcal{A}$ and all $g \in G$) and that $\tilde{G}$ is commutative. He then gives sufficient conditions for estimators δ of θ to be inadmissible. He does this by explicitly constructing, for each $x \in \mathcal{X}$, a closed, convex subset Θ_x of Θ such that

$$[P_\theta(\delta(X) \notin \Theta_X) > 0 \text{ for some } \theta \in \Theta] \implies [\delta \text{ is inadmissible}].$$

Further, he shows that the estimator δ_o defined, for each x, as the projection of $\delta(x)$ unto Θ_x dominates δ. Of course, when $\Theta_x = \Theta$ for all $x \in \mathcal{X}$ nothing can be concluded concerning the (in)admissibility of δ.

The sets Θ_x are defined as follows. Let $G = (g_1, \ldots, g_p)$ and let, for $x \in \mathcal{X}$ and $\theta \in \Theta$ (see Moors, 1985, p. 43),

$$\alpha(x, \bar{g}_j(\theta)) = \frac{p_{\bar{g}_j(\theta)}(x)}{S(x; \theta)}, \qquad j = 1, \ldots, p,$$

when $S(x; \theta) = \sum_{i=1}^{p} p_{\bar{g}_i(\theta)}(x) > 0$. Further, let

$$t_x(\theta) = \begin{cases} \sum_{i=1}^{p} \alpha(x, \bar{g}_i(\theta)) \tilde{g}_i(\theta) \text{ when } S(x; \theta) > 0 \\ \\ \theta \qquad\qquad\qquad\qquad \text{when } S(x; \theta) = 0. \end{cases}$$

Then Θ_x is the convex closure of the range of $t_x(\theta)$.

Moors gives examples of estimation problems in randomized response models, in the simple linear regression problem with a restriction on the slope parameter as well as in the estimation of a restricted binomial parameter. To illustrate Moors' results, let $X \sim \text{Bin}(1, \theta)$ with $\Theta = [m, 1-m]$ for some known $m \in (0, 1/2)$. This problem is invariant with respect to the group $G = (g_1, g_2)$, where $g_1(x) = x$ and $g_2(x) = 1 - x$. The sets Θ_x are given by

$$\Theta_x = \begin{cases} [\dfrac{1 - \phi}{2}, 1/2] \text{ when } x = 0 \\ \\ [1/2, \dfrac{1 + \phi}{2}] \text{ when } x = 1, \end{cases}$$

where $\phi = (2m - 1)^2$. This means that any estimator δ for which

$$\delta(0) \notin [2m(1 - m), 1/2] \quad \text{or} \quad \delta(1) \notin [1/2, 1 - 2m(1 - m)] \qquad (3.6)$$

is inadmissible and dominated by its projection on the interval $[2m(1-m), 1 - 2m(1 - m)]$. A particular case of such an inadmissible estimator is the MLE, which is given by

$$\text{MLE}(X) = \begin{cases} m \qquad \text{when } X = 0 \\ 1 - m \text{ when } X = 1. \end{cases}$$

Of course, this inadmissibility result, but not the dominators, can be obtained from Brown's (1986) Theorem 4.23. Another example where Moors (1985, p. 94-96) shows inadmissibility for a large class of estimators (with dominators) is the case of Simmons' unrelated-question randomized-response model (see Greenberg, Abul-Ela, Simmons and Horvitz, 1969). In this case X_1 and X_2 are independent with $X_i \sim \text{Bin}(n_i, \theta_i)$, $i = 1, 2$ and

$$\theta_i = (1 - m_i)\pi_A + m_i \pi_Y, \qquad i = 1, 2 \qquad (3.7)$$

for known $m_i \in (0, 1), i = 1, 2$. The vector (π_A, π_Y) is to be estimated under the restrictions imposed by (3.7). In this case Brown does not apply, because the parameter space is not a closed rectanglar subset of the natural parameter space $(0, 1) \times (0, 1)$.

As a third example, let $X \sim \mathcal{N}(\theta, 1)$ with $\theta \in [-m, m]$. This problem is invariant with respect to the group $G = (g_1, g_2)$, where $g_1(x) = x$ and $g_2(x) = -x$. The sets Θ_x are given by

$$\Theta_x = \{\theta \mid -m \tanh(m|x|) < \theta < m \tanh(m|x|)\}$$

So, any δ satisfying

$$P_\theta(\delta(X) \in \Theta_X) < 1 \quad \text{for some } \theta \in [-m, m]$$

is inadmissible and dominated by its projection unto Θ_X. The MLE is of course such an estimator.

Kumar and Sharma (1992) show that Moors' results hold under weaker conditions on the group G than those of Moors. It does not need to be finite – locally compact is sufficient. Moors and van Houwelingen (1993) further extend Moors' (1981, 1985) results. They consider the linear model $X = Z\theta + \varepsilon$, where $X = (X_1, \ldots, X_n)'$ with the X_i independent normal with known variances and Z an $n \times k$ known matrix and show that $\tilde{G}$ does not need to be commutative and that g does not need to be measure-preserving. The restrictions on the unknown $k \times 1$ vector of parameters θ are either of the form $m_1 \leq A\theta \leq m_2$ for known $k \times 1$ vectors m_1 and m_2 and known $k \times k$ matrix A and with the inequalities componentwise, or of the form $\theta'B\theta \leq m$ for a known diagonal $k \times k$ matrix B and a known positive m. More on these Moors and van Houwelingen results can be found in Chapter 6, Section 6.3.

3.3 Results of Charras and van Eeden

Charras and van Eeden (1991a,b, 1994) consider, like Moors (1981, 1985), the general case as described at the beginning of Chapter 2. Their loss is squared error and Θ is closed and convex. These authors give sufficient conditions for inadmissibility of so-called "boundary estimators" (terminology introduced by Moors (1985)), i.e., estimators satisfying

$$\text{there exists } A_o \in \mathcal{A} \text{ such that } \begin{cases} i) \ \delta(A_o) \subset B(\Theta) \\ \\ ii) \ \nu(A_o) > 0, \end{cases} \tag{3.8}$$

where $B(\Theta)$ is the boundary of Θ and $\delta(A_o) = \{\delta(x) \mid x \in A_o\}$.

The first of their results (Charras and van Eeden, 1991a) holds for compact parameter spaces. They use Wald's (1950, Theorem 5.9) condition

$$\lim_{\theta \to \theta_o} \int_{\mathcal{X}} |p_\theta(x) - p_{\theta_o}(x)| \, d\nu(x) = 0 \quad \text{for all } \theta_o \in \Theta. \tag{3.9}$$

Under this condition, admissible estimators are Bayes when the parameter space is compact. So, sufficient conditions for an estimator δ to be non-Bayes are then sufficient conditions for δ to be inadmissible.

We first illustrate their results by means of some examples. Let X have a logistic distribution with density

$$p_\theta(x) = \frac{\exp(-(x - \theta))}{(1 + \exp(-(x - \theta)))^2} \quad -\infty < x < \infty$$

and let $\theta \in [m_1, m_2]$ with $-\infty < m_1 < m_2 < \infty$. Let δ_1 be an estimator of θ such that there exists measurable sets A_1 and A_2 with $\nu(A_i) > 0$, $i = 1, 2$, $A_1 \subset (-\infty, m_1)$, $A_2 \subset (m_2, \infty)$, $\delta_1(A_1) = m_1$ and $\delta_1(A_2) = m_2$. The MLE is an example of such an estimator. If this δ_1 is Bayes with respect to some prior on the interval $[m_1, m_2]$, the posterior distribution has support $\{m_1\}$ when $x \in A_1$ and support $\{m_2\}$ when $x \in A_2$. But the posterior and the prior distributions have the same support, so this δ_1 can not be Bayes. Another estimator of the mean of a logistic distribution which is inadmissible is given by

$$\delta_2(x) = \begin{cases} m_1 & \text{when } x \leq m_1 \\ \delta(x) \in [m_1 + \varepsilon, m_2 - \varepsilon] & \text{when } x > m_1, \end{cases}$$

where $0 < \varepsilon < (m_2 - m_1)/2$. For this case, note that, for $x \in A_1$, the support of the posterior, and therefore of the prior, is $\{m_1\}$. But this contradicts the fact that, for $x \in A_2$, $\delta_2 \geq m_1 + \varepsilon$. These results for the logistic distribution can not be obtained from Brown's (1986) Theorem 4.23 nor from his Theorem 4.16 because the logistic is not an exponential-family distribution.

As a second example of the results of Charras and van Eeden (1991a), let $X \sim$ Bin(n, θ) with $0 \leq \theta \leq m$ for some $m \in (0, 1)$. Let $n \geq 2$ and let δ_3 be the following boundary estimator

$$\begin{cases} \delta_3(0) = 0 \ < \delta_3(1) \\ \delta_3(x) = m \text{ for } x \geq 2. \end{cases} \tag{3.10}$$

This estimator is the MLE of θ when $(1 \leq nm \leq 2, n\delta_3(1) = 1)$ or $(nm < 1, \delta_3(1) = m)$ and, as already noted above, Charras and van Eeden (1991a) show that this MLE is admissible. In the more general case where δ_3 is not necessarily the MLE, it is easy to see that (3.10) is (non-unique) Bayes with respect to the prior with support $\{0\}$, which tells us nothing about its admissibility. But Charras and van Eeden (1991a, p. 128) show that this δ_3 is admissible, which shows that not every boundary estimator is inadmissible.

The above two examples are special cases of the following general results of Charras and van Eeden (1991a, Theorems 3.1 – 3.3). Let δ be a boundary estimator and let, for x such that $\delta(x) \in B(\Theta)$,

$$G(x) = \text{face}(\delta(x)) \cup \overline{\{\theta \in \Theta | p_\theta(x) = 0\}} \qquad (3.11)$$

and let

$$\Theta^* = \{\theta \in \Theta | \theta \in G(x) \text{ for } \nu\text{-almost all } x \in A_o\}, \qquad (3.12)$$

where A_o is defined in (3.8) and, for $\theta \in B(\Theta)$, face(θ) is the intersection of Θ with the intersection of all its tangent planes at θ. Then, if δ is Bayes the support of its prior is contained in Θ^*. From this it follows that a boundary estimator for which Θ^* is empty is inadmissible. The above estimator δ_1 of the logistic mean is an example of such a boundary estimator.

Further, Charras and van Eeden (1991a) prove that, if Θ^* is not empty and there exists $A^* \in \mathcal{A}$ such that

$$
\begin{cases}
i) & \nu(A^*) > 0 \\
\\
ii) & \delta(A^*) \cap \overline{co}(\Theta^*) \text{ is empty} \qquad (3.13) \\
\\
iii) & p_\theta(x) > 0, \text{ for all } x \in A^*, \text{ and all } \theta \in \Theta^*,
\end{cases}
$$

where $\overline{co}(\Theta^*)$ is the convex hull of Θ^*, then δ is again inadmissible. The estimator δ_2 in the above logistic example satisfies these conditions.

A second result of Charras and van Eeden can be found in their (1991b) paper. It is concerned with inadmissibility of boundary estimators when $\Theta \subset R^k$ is closed and convex, but not necessarily compact and $k \geq 2$. The authors put conditions on the shape of the boundary of Θ as well as some (rather weak) ones on the family of densities. Examples of when these conditions are satisfied are the case where Θ is strictly convex, $X_1, \ldots, X_k$ are independent and, for $i = 1, \ldots, k$, X_i has density $p_{\theta_i}(x)$ which is positive for each $x \in \mathcal{X}$ and $\theta \in \Theta$ and bounded on Θ for each $x \in \mathcal{X}$. The authors give conditions on the estimators under which they are inadmissible. This result can be used, e.g., to show that, when $k = 2$ and the X_i have logistic distributions with means θ_i, the MLE of $\theta = (\theta_1, \theta_2)$ when these parameters are restricted to $\Theta = \{\theta \mid \theta_2 \geq c \theta_1^2\}$ for a known $c > 0$ is inadmissible.

A third result of Charras and van Eeden, to be found in their 1994 paper, is concerned with the inadmissibility of boundary estimators when $\Theta = \{\theta \mid \theta \geq m_1\}$ for some $m_1 < \infty$. The authors suppose that the estimators δ are such that there exists sets A_1 and A_2 in $\mathcal{A}$ and an $m_2 > m_1$ with

$$i) \; \delta(x) = m_1 \qquad\qquad \text{when } x \in A_1$$

$$ii) \; \delta(x) \geq m_2 \qquad\qquad \text{when } x \in A_2 \qquad\qquad (3.14)$$

$$iii) \; 0 < \nu(A_i) < \infty, i = 1, 2.$$

They show that their further conditions, which are non-trivial conditions on the family of distributions, are satisfied for gamma and logistic scale families. That these conditions are also satisfied for a scale family of F distributions was proved by van Eeden and Zidek (1994a,b). Examples of estimators which satisfy (3.14) for these three families of distributions are the MLE of θ^p, $p > 0$.

3.4 Dominators

In this section dominators are presented for estimators which are inadmissible but satisfy (2.3). The dominators do not necessarily satisfy (2.3).

As already said above, for each one of Moors' (1981, 1985) inadmissible estimators, dominators are known, because Moors proves inadmissibility by constructing dominators – and his dominators satisfy (2.3).

Charras and van Eeden (1991a), using squared-error loss, give two classes of dominators for some of their inadmissible boundary estimators. They suppose that $\Theta = \{\theta \mid m_1 \leq \theta \leq m_2\}$, $-\infty < m_1 < m_2 < \infty$, and that δ is such that both $P_\theta(\delta(X) = m_1)$ and $P_\theta(\delta(X) = m_2)$ are positive for all $\theta \in [m_1, m_2]$. One of these classes of dominators of an inadmissible estimator δ consists of estimators of the form

$$\delta^*(X) = \begin{cases} m_1 + \varepsilon_1 & \text{when } \delta(X) = m_1 \\ \delta(X) & \text{when } m_1 < \delta(X) < m_2 \\ m_2 - \varepsilon_2 & \text{when } \delta(X) = m_2, \end{cases} \qquad (3.15)$$

where $\varepsilon_1 + \varepsilon_2 < m_2 - m_1$. This dominator was proposed by a referee of their (1991a) paper and his proof goes as follows. Let, for $t \in [m_1, m_2]$,

$$p_1(t) = \min\{P_\theta(\delta(X) = t) \mid m_1 \leq \theta \leq m_2\}$$
$$p_2(t) = \max\{P_\theta(\delta(X) = t) \mid m_1 \leq \theta \leq m_2\}.$$

Then the following three cases are possible:

$$\varepsilon_1 p_2(m_1) < \varepsilon_2 p_1(m_2);$$
$$\varepsilon_1 p_2(m_1) \geq \varepsilon_2 p_1(m_2) \text{ and } \varepsilon_1 p_1(m_1) \leq \varepsilon_2 p_2(m_2);$$
$$\varepsilon_1 p_1(m_1) > \varepsilon_2 p_2(m_2)$$

and, in each of these three cases, it can easily be seen that there exist ε_1 and ε_2 such that the estimator (3.15) dominates δ. Note that δ^* is, in general, not monotone and, if not monotone, inadmisible when X has an exponential-family distribution with mean θ. (see Brown, 1986, Theorem 4.23).

The other class of dominators of van Eeden and Charras (1991a) is of the form

$$\delta^{**}(X) = \begin{cases} m_1' & \text{when } \delta(X) < m_1' \\ \delta(X) & \text{when } m_1' \leq \delta(X) \leq m_2' \\ m_2' & \text{when } \delta(X) > m_2', \end{cases} \tag{3.16}$$

where $m_1 < m_1' < m_2' < m_2$. The proof that there exist m_1' and m_2' for which δ^{**} dominates δ can be found in Charras (1979). His proof goes as follows: Let $a(t)$ and $b(t)$ be two functions defined on $[0, 1]$ with $a(0) = m_1, b(0) = m_2, a(1) = b(1)$, $a(t)$ non-decreasing and $b(t)$ non-increasing. Then under regularity conditions on $a(t)$, $b(t)$ and the distribution of δ, he shows that there exist $t_o \in (0, 1)$ such that δ^{**} with $m_1' = m_1 + a(t_o)$ and $m_2' = m_2 - b(t_o)$ dominates δ.

Note that the estimator (3.16) is, often, the MLE of θ for the parameter space $[m_1', m_2']$.

As an example of the dominators (3.15), consider the case where $X \sim \text{Bin}(n, \theta)$, $m \leq \theta \leq 1 - m$, $\varepsilon_1 = \varepsilon_2 = \varepsilon$ and $n = 1$ or 2. Then the Charras–van Eeden (1991a) conditions on m and ε to dominate the MLE become

$$0 < \varepsilon \leq \min\left(2m(1 - 2m), \frac{1 - 2m}{2}\right) \qquad \text{when } n = 1 \tag{3.17}$$

and

$$0 < \varepsilon \leq \min\left(\frac{2m^2(1 - 2m)}{2m^2 - 2m + 1}, \frac{1 - 2m}{2}\right) \qquad \text{when } n = 2. \tag{3.18}$$

When $m \geq 1/4$ and $n = 1$, the condition (3.17) gives $m < \delta^*(0) \leq 1/2$, implying that every symmetric estimator dominates the MLE. As already mentioned above, Hodges and Lehmann (1950) proved this for the case where $m = 1/3$. When $m < 1/4$ the condition (3.17) gives $m < \delta^*(0) < m(3 - 4m)$ implying that, as expected, the MLE can not be improved upon very much when m is close to zero. These classes of Charras–van Eeden dominators of the MLE for this binomial example are not those of Moors. When $n = 1$ and $m < 1/4$, e.g., Moors' dominator, δ_M say, satisfies (see (3.6)) $m < \delta_M(0) \leq 2m(1-2m)$. Given that $2m(1 - 2m) < m(3 - 4m)$ when $0 < m < 1/2$, the Charras–van Eeden class is larger than than the Moors class.

Still for this binomial example with $n = 1$ or $n = 2$ and with $\varepsilon_1 = \varepsilon_2 = \varepsilon > 0$ satisfying (3.17) when $n = 1$ and (3.18) when $n = 2$, Marchand and MacGibbon (2000) show, in their Theorem 4.2, that the dominators (3.15) are, for squared-error loss, unique Bayes with respect to a symmetric prior on $[m, 1 - m]$. This implies that they are admissible. Further, Marchand and MacGibbon (2000, p. 144) state that dominators of the form (3.15) may not be admissible, but they do not relate this statement to the consequences of their Theorem 4.2. What is happening here is that the dominators (3.15) are not necessarily monotone and, when they are not monotone, they are inadmissible in the exponential-family setting of Brown (1986, Theorem 4.23). But for the above binomial cases – where Brown's theorem does apply – the dominators of the form (3.15) are monotone, so there is no contradiction.

Remark 3.3. Marchand and MacGibbon (2000, Theorem 4.2) forgot to mention a necessary condition on (m, ε), namely, $m + \varepsilon \le 1/2$.

More about dominators when $X \sim \text{Bin}(n, \theta)$ with $m \le \theta \le 1 - m$ for a known $m \in (0, 1/2)$ can be found in Perron (2003). He uses squared-error loss and gives sufficient conditions for an estimator to dominate the MLE. For Moors' (1985) dominator of the MLE Perron shows that it is the Bayes estimator with respect to a symmetric prior om $\{m, 1 - m\}$ if and only if $1 - 2m$ is $\le 1/\sqrt{n}$ when n is odd and $\le 1/(\sqrt{n-1})$ when n is even. For the Charras–van Eeden dominators (3.16) of the MLE Perron gives an algorithm for finding (m'_1, m'_2) when $m'_1 + m'_2 = 1$. He also proposes a new estimator, namely the Bayes estimator with respect to a prior proportional to $(\theta(1-\theta))^{-1}$ and shows that it dominates the MLE for some n, but not for all. For Perron's graphical comparisons between these estimators, see Section 3.6.

Remark 3.4. Ali and Woo (1998) also look at the case where θ is to be estimated with squared-error loss when $X \sim \text{Bin}(n, \theta)$ with $m \le \theta \le 1 - m$ for a known $m \in (0, 1/2)$. They have four estimators, namely, $\check{\theta} = (1-2m)X/n+m$, the MLE $\hat{\theta}$ and two Bayes estimators each with a truncated beta prior, $\tilde{\theta}$ with squared-error loss and θ^* with a linex loss function.
Some of their results are incorrect. For instance, their statement that the MSE of $\check{\theta}$ is strictly increasing in m as m increases from $1/(2(1+\sqrt{n})$ to $1/2$.
Further, they make comparisons between the risk functions only numerically for very small numbers of values of n, θ, m and the parameters of the priors, but conclude, e.g., (on the basis of one such set of values) that $\hat{\theta}$ is worse than $\check{\theta}$ in the sense of MSE.
And finally, in comparing $\tilde{\theta}$ with θ^*, they use squared-error loss for $\tilde{\theta}$ and linex loss $L(\theta^*, \theta) = e^{\theta^* - \theta} - (\theta^* - \theta) - 1$ for θ^*, i.e., the loss function used for the Bayes estimator θ^*. But the same Bayes estimator would have been obtained from the loss function $cL(\theta^*, \theta)$ for any positive c with a different comparison result.

Other cases where dominators are known are, for example,

1) The case studied by Sackrowitz and Strawderman (1974) where $X_1, \ldots, X_k$, $k \geq 2$, are independent with, for $i = 1, \ldots, k$, $X_i \sim \text{Bin}(n_i, \theta_i)$, $\Theta = \{\theta \mid \theta_1 \leq \ldots \leq \theta_k\}$ and the loss function is of the form $\sum_{i=1}^{k} W(|d_i - \theta_i|)$ with W strictly convex. Sackrowitz (1982) obtains dominators for the inadmissible MLE's;

2) The case of the lower-bounded mean of a normal distribution, as well as the case of a lower-bounded scale parameter of a gamma distribution. For each of these cases Shao and Strawderman (1996a,b) give a class dominators when the loss is squared error. For the normal-mean case they dominate the MLE and say they believe, but have not proved, that some of their dominators are admissible. For the gamma-scale case they dominate truncated linear estimators. Among their results is the interesting case of an estimator which is admissible for the non-restricted parameter space, takes values in the restricted space only and is inadmissible for the restricted parameter space;

3) Shao and Strawderman (1994), among their already earlier mentioned power series distribution results, give dominators for squared-error loss for the MLE of the mean when the parameter space is restricted;

4) And lastly in this list of results on dominators, the case of the multivariate normal mean (and, more generally, the multivariate location problem). For $X \sim \mathcal{N}_k(\theta, I)$ it is well known from Stein (1956) that, for squared-error loss, X is inadmissible as an estimator of θ when $k \geq 3$ and $\Theta = R^k$. Dominators are known, the James–Stein (1961) estimator, for instance. For restricted spaces (and I am restricting myself here, as in the rest of this monograph, to closed convex subsets of R^k) quite a number of dominators have been obtained for various inadmissible estimators.

Chang (1991) looks at $X = (X_{1,1}, X_{1,2}, \ldots, X_{k,1}, X_{k,2}) \sim \mathcal{N}_{2k}(\theta, I)$ where

$$\theta = (\theta_{1,1}, \theta_{1,2}, \ldots, \theta_{k,1}, \theta_{k,2})$$

satisfies $\theta_{i,1} \leq \theta_{i,2}$, $i = 1, \ldots, k$. He estimates θ under squared-error loss. For this case it is known that, for estimating $\theta_i = (\theta_{i,1}, \theta_{i,2})$ based on $X_i = (X_{i,1}, X_{i,2})$ alone, the Pitman estimator (i.e., the Bayes estimator with respect to the uniform prior on $\Theta_i = \{(\theta_{i,1}, \theta_{i,2}) \mid \theta_{i,1} \leq \theta_{i,2}\}$) is admissible (see Chapter 4, Section 4.4). Chang (1991) shows that, analogous to Stein's (1956) result, the vector of these Pitman estimators is inadmissible for estimating θ when $k \geq 2$. One of his two classes of James–Stein-type dominators is given by $(1 - c/S)\delta_P(X)$, where $\delta_P(X) = (\delta_{P,1}(X_1), \ldots, \delta_{P,k}(X_k))$ is the vector of Pitman estimators, $S = \sum_{i=1}^{k}(\delta_{P,i}(X_i))^2$ and $0 < c \leq 4(k-1)$. He has a second class of dominators for the case where $k \geq 3$.

Chang (1982) looks at the case where $\Theta = \{\theta \mid \theta_i \geq 0, i = 1, \ldots, k\}$ with squared error loss and $X \sim \mathcal{N}_k(\theta, I)$. Here, as we have seen, the MLE of θ_i based on X_i alone is inadmissible (and minimax) for estimating θ_i and, as will be seen in Chapter 4, Section 4.3, Katz's estimator $\delta_{K,i}(X_i) = X_i + \phi(X_i)/\Phi(X_i)$ is admissible for estimating θ_i, where ϕ and Φ are, respectively, the standard normal density and distribution function. Chang (1982) shows that, again analogous to Stein (1956), the vector of Katz estimators is inadmissible for estimating θ when $k \geq 3$. One of his classes of James–Stein-type dominators is given by $\delta(X) = (\delta_1(X), \ldots, \delta_k(X))$, with, for $i = 1, \ldots, k$,

$$\delta_i(X) = \begin{cases} \delta_{K,i}(X_i) - \dfrac{cX_i}{\sum_{j=1}^{k} X_j^2} & \text{when } X_l \geq 0,\, l = 1, \ldots, k \\ \delta_{K,i}(X_i) & \text{otherwise,} \end{cases}$$

with $0 < c < 2(k-2)$. A second class is given by

$$\delta_i^*(X) = \delta_{K,i}(X_i) - \frac{c\delta_{K,i}(X_i)}{\sum_{j=1}^{k} \delta_{K,j}^2(X_j)}, \qquad i = 1, \ldots, k,$$

again with $0 < c < 2(k-2)$. Chang (1982) also shows that replacing, in these dominators, $\delta_{K,i}$ by the MLE of θ_i based on X_i alone gives a dominator of the vector of MLEs. Chang (1981), again using squared-error loss, dominates the MLE of θ when $X \sim \mathcal{N}_k(\theta, I)$, $\Theta = \{\theta \mid A\theta \geq 0\}$ for a known $p \times k$ matrix A of rank k, $p \leq k$, $k \geq 3$ and $A\theta \geq 0$ denotes the componentwise inequalities. An example of his dominators is the James–Stein-type estimator

$$\delta(X) = \begin{cases} \left(1 - \dfrac{c}{\sum_{i=1}^{k} X_i^2}\right) X & \text{when } AX \geq 0 \\ \text{the MLE of } \theta & \text{otherwise,} \end{cases} \qquad (3.19)$$

where $0 < c < 2(k-2)$.

Sengupta and Sen (1991) consider $X \sim N_k(\theta, \Sigma)$ with Σ positive definite, known or unknown, $k \geq 3$, θ restricted to a closed, convex subset of R^k and loss function $(d - \theta)'\Sigma^{-1}(d - \theta)$. In particular, they consider the suborthant model where $\Theta = \Theta_{k_1}^+ = \{\theta \mid \theta_i \geq 0, i = 1, \ldots, k_1\}$ for some $k_1 \in \{1, \ldots, k\}$. They define restricted Stein–rule MLEs $\hat{\theta}_{RS}$ and positive restricted Stein–rule MLEs $\hat{\theta}_{PRS}$ of θ for $\theta \in \Theta_{k_1}^+$. One of their results states that $\hat{\theta}_{PRS}$ dominates $\hat{\theta}_{RS}$ which dominates the (restricted) MLE. On the relationship of their work with that of Chang (1981, 1982) Sengupta and Sen say that the full impact of shrinkage has not been incorporated in the estimators considered by Chang and they give dominators of Chang's (1982) dominators.

Still for this multivariate location problem, Kuriki and Takemura (2000, Section 3.3) weaken the conditions on Θ for one of Sengupta and Sen's (1991) domination results and Ouassou and Strawderman (2002) and Fourdrinier, Ouassou and Strawderman (2003) generalize and extend the Chang (1981, 1982) and Sengupta and Sen (1991) results to spherically symmetric distributions.

In none of the above-quoted papers concerning multivariate location problems do the authors mention the question of whether their dominators satisfy (2.3). Those of Chang (1981, 1982, 1991) do not, nor do Sengupta and Sen's (1991) dominators of Chang's (1982) dominators. As far as the other results are concerned, I do not know whether their dominators satisfy (2.3). But (see Chapter 2), given that the class of estimators $\mathcal{D}$ is essentially complete in the class $\mathcal{D}_o$ with respect to Θ, an estimator which does not satisfy (2.3) can be replaced by one which does satisfy (2.3) and dominates it on Θ.

Sengupta and Sen (1991) give two examples of models which can be reduced to their suborthant model: namely, the ordered model where $X_{i,j} \sim^{ind} \mathcal{N}(\theta_i, \sigma^2)$, $j = 1, \ldots, n_i$, $i = 1, \ldots, k$ with $\Theta = \{\theta \mid \theta_1 \leq \ldots, \theta_r\}$ for some $r \in \{2, \ldots, k\}$ and the two-way layout where $X_{i,j} = \mu_i + \theta_j + \varepsilon_{i,j}$, $j = 1, \ldots, n$, $i = 1, \ldots, n$, $\varepsilon_{i,j} \sim^{ind} \mathcal{N}(0, \sigma^2)$ and the same Θ as for the ordered alternative model. They also note that each of these are particular cases of some linear model. More on (normal) linear models with restricted parameter spaces can be found in Chapter 6.

Remark 3.5. Some of the James–Stein dominators presented above only hold for the case where $\sigma^2 = 1$. Of course, when σ^2 is known, one can, without loss of generality, take it to be $= 1$. However, this might lead to mistakes if somebody whose σ^2 is $\neq 1$ uses the results without realizing that $\sigma^2 = 1$ is assumed in deriving the results. So, in (3.19), e.g., I would prefer to write $c\sigma^2$ (with $0 < c\sigma^2 < 2(k-2)$) instead of c.

A multivariate location problem with constraints on the norm can be found in Marchand and Perron (2001). Let $X_1, \ldots, X_k$ be independent with $X_i \sim \mathcal{N}(\theta_i, 1)$, $i = 1, \ldots, k$, and $\sum_{i=1}^{k} \theta_i^2 \leq m^2$ for a known, positive constant m. Marchand and Perron (2001) consider the problem of finding dominators for the MLE of θ for squared-error loss. One of their results says that the Bayes estimator with respect to the uniform prior on the boundary of Θ dominates the MLE when $m \leq \sqrt{k}$. Casella and Strawderman (1981) prove this for $k = 1$. A generalization of their results to spherical symmetric distributions, in particular to the multivariate student distribution, can be found in Marchand and Perron (2005).

Remark 3.6. Chang (1982) quotes and uses Katz's (1961) unproven results concerning admissible estimators for the lower-bounded scale parameter of a gamma distribution, as well as those for a lower-bounded mean of a Poisson distribution. Katz's proofs are incorrect for these cases (see Remark 4.4).

Remark 3.7. The three papers by Charras and van Eeden (1991a,b, 1994), as well as a fourth one (Charras and van Eeden, 1992) not discussed in this monograph, are based on results obtained by Charras in his 1979 PhD thesis.

3.5 Universal domination and universal admissibility

Hwang's (1985) universal domination and universal admissibility have been applied, mostly with $D = I$, to various problems in restricted-parameter-space estimation. An example of the kind of results that have been obtained can be found in Cohen and Kushary (1998). They show the universal admissibility of the MLE of θ for, e.g., (i) the case where $X \sim \mathcal{N}_k(\theta, I)$, $k \geq 2$, θ is restricted to a polyhedral cone and the class of estimators consists of those which are nonrandomized and continuous; (ii) the case where X has a discrete (possibly multivariate) distribution and the MLE is unique and takes a finite number of values; (iii) the bounded normal mean case with known variance. In this last case, the universal admissibility of the MLE also follows from Hwang's (1985) results and the fact that the MLE of θ is admissible for absolute error loss, as was shown by Iwasa and Moritani (1997).

Further results on universal domination and universal admissibility in restricted parameter spaces are presented in later chapters.

3.6 Discussion and open problems

As seen above, although the three main results on inadmissibility of estimators $\delta \in \mathcal{D}$ – those of Brown, of Moors and of Charras and van Eeden – cover quite a lot of situations, many questions remain open. Moreover, almost all known results are only for squared-error loss.

What would be a great help in filling the gaps in our knowledge would be an extension of Brown's (1986) Theorem 4.23 to k ($k \geq 1$) dimensions and Θ an arbitrary closed convex subset of N^k. That would solve many inadmissibility questions for the exponential family. An example of an open problem of this kind is the question of whether, for squared-error loss, the MLE of the largest of $k \geq 3$ ordered scale parameters of a gamma distributions is admissible. Another open problem for restricted exponential-family parameters is the one mentioned in Hoferkamp and Peddada (2002). They have $X \sim \mathcal{N}_k(\theta, \Sigma)$ with Σ diagonal, $\theta_1 \leq \ldots \leq \theta_k$ and $\sigma_1^2 \leq \ldots \leq \sigma_k^2$ and note that it seems to be unknown whether the MLE of $(\theta_1, \ldots, \theta_k, \sigma_1^2, \ldots, \sigma_k^2)$ is admissible. Another

very useful addition would be a version of the results of Charras and van Eeden (1991a,b, 1994) with weaker conditions on the parameter space.

A problem not touched upon in this chapter is the question of "how inadmissible" the inadmissible estimators are, i.e., how large (or small) is the improvement of a dominator (preferably an admissible dominator) over the estimator
it dominates. For some models numerical results have been obtained. Examples of such results can be found in Shao and Strawderman (1996b), as well
as in Perron (2003). Shao–Strawderman's tables on the improvement of their
dominators over the MLE for a lower-bounded normal mean give a maximum
improvement of $\approx .0044\sigma$. This is not much of an improvement over the risk
function of the MLE, which varies from $\sigma/2$ to σ over the interval $[0, \infty)$.
For the binomial case where $X \sim \text{Bin}(n, \theta)$ with $\theta \in [m, 1 - m]$, Perron gives
graphs of the risk functions of the MLE, of its Charras–van Eeden dominator
(3.16), of its Moors' dominators and of (see Section 3.4) Perron's Bayes estimator with respect to a prior proportional to $(\theta(1 - \theta))^{-1}$. From these graphs
one can get an (approximate) idea of the absolute as well as the relative improvements over the MLE of these dominators. For $n = 10$ (for which Perron's
Bayes estimator does dominate the MLE) and $m = .40$, e.g., the maximum
improvements vary, for the three dominators, from $\approx .005$ to $\approx .007$. The corresponding relative improvements are (again approximately) 75%, 68% and
95% for, respectively, the Charras–van Eeden, the Moors and the Perron dominator. For $m = .4$ and $n = 25$ and 100, e.g., Perron finds not only that his
Bayes estimator dominates the MLE on almost the whole parameter space
(in fact on $\approx (.42 \le \theta \le .58)$), but that, for these same θ's, it dominates
both Moors' and Charras and van Eeden's dominator (3.16) of the MLE. (In
the caption of Perron's Figure 3, the $n = 10$ should be $n = 1000$). Further,
for $n = 1000$, there is, essentially, no difference between the risk functions of
the MLE, Moors's dominator and Charras and van Eeden's dominator (3.16),
while Perron's Bayes estimator dominates these three on $\approx (.41 \le \theta \le .59)$.
So, for the binomial case with $m \le \theta \le 1 - m$, much "better" dominators are
available than for the lower bounded normal mean case.

More such numerical results are presented in Chapter 7, Section 7.2 .

Finding dominators is clearly a very difficult problem. Look at the time it took
– from the early 1960s until 1996 – to find a dominator for the MLE of a lower-
bounded normal mean. And look at how few cases have been solved, other
than under the (rather restrictive) conditions of Moors (1981, 1985). Further,
very few admissible dominators have been found. One example, estimating
$\theta \in [m, 1 - m]$ with squared-error loss when $X \sim \text{Bin}(n, \theta)$ with $n = 1$ or $= 2$,
is mentioned above. More examples of admissible estimators (not necessarily
dominators) can be found in later chapters, in particular in Chapter 4, Section
4.4, where (admissible) minimax estimators are presented. One very special
case where an admissible dominator has been obtained can be found in Parsian

and Sanjari Farsipour (1997). They consider estimating θ on the basis of a sample $X_1, \ldots, X_n$ from a distribution with density $e^{-(x-\theta)/\sigma}/\sigma$, $x > \theta$, with known σ and the restriction $\theta \leq 0$. They use the linex loss function $L(d, \theta) = e^{a(d-\theta)} - a(d-\theta) - 1$ for a known nonzero constant $a < n/\sigma$. For the unrestricted case, the Pitman estimator is given by (see Parsian, Sanjari Farsipour and Nematollahi, 1993) $\min(X_1, \ldots, X_n) - \log(n/(n-a\sigma))/a$ and Parsian and Sanjari Farsipour (1997) show that replacing $\min(X_1, \ldots, X_n)$ by $\min(c, \min(X_1, \ldots, X_n))$ gives the Pitman estimator for the restricted θ. This estimator clearly satisfies (2.3) and the authors show it to be admissible. Whether this estimator is minimax seems to be unknown (see Chapter 4, Section 4.3). Note that neither estimator is scale-equivariant, unless $a\sigma = a^*$ for some nonzero constant $a^* < n$.

4

Minimax estimators and their admissibility

In this chapter, results on minimax estimation of θ in restricted parameter spaces are given for the case where the problem does not contain any nuisance parameters, i.e., (in the notation of Chapter 2): $K = M$, $\Omega_o = \Theta_o$ and $\Omega = \Theta$ and we change the notation M for the number of parameters to be estimated to k. The results are presented in the sections 4.2, 4.3 and 4.4, where, respectively, Θ is bounded, Θ is not bounded with $k = 1$ and Θ is not bounded with $k > 1$. In the introductory Section 4.1 some known results which are useful for solving minimax problems are given. Most of these results apply to restricted as well as to unrestricted parameter spaces. The last section of this chapter contains some comments and open problems.

4.1 Some helpful results

The lemmas 4.1 and 4.2 can be found, e.g., in Lehmann (1983, pp. 249 and 256), in Lehmann and Casella (1998, pp. 310 and 316) and in Berger (1985, p. 350):

Lemma 4.1 *Suppose that π_o is a prior for $\theta \in \Theta$ such that, for a given loss function, the Bayes risk r_{π_o} of the Bayes rule δ_{π_o} satisfies*

$$r_{\pi_o} = \sup_{\theta \in \Theta} R(\delta_{\pi_o}, \theta).$$

Then, for that loss function, δ_{π_o} is minimax and, if δ_{π_o} is unique Bayes, it is unique minimax. Finally, the prior π_o is least favourable, i.e., $r_{\pi_o} \geq r_\pi$ for all priors π on Θ.

The conditions of Lemma 4.1 imply that the statistical problem has a value, i.e.,

$$\sup_\pi \inf_\delta r_\pi(\delta) = \inf_\delta \sup_\theta R(\delta, \theta), \tag{4.1}$$

or, equivalently

$$r_{\pi_o} = \inf_{\delta} \sup_{\theta} R(\delta, \theta),$$

which says that the maximum Bayes risk equals the minimax value.

Lemma 4.2 *Let π_n, $n = 1, 2, \ldots$, be a sequence of priors for $\theta \in \Theta$ and let, for a given loss function, $r_{\pi_n} \to r$ as $n \to \infty$. Further suppose that there exists an estimator δ such that*

$$\sup_{\theta \in \Theta} R(\delta, \theta) = r.$$

Then, for that loss function, δ is minimax and the sequence π_n is least favourable, i.e., it satisfies

$$r \geq \int_{\Theta} R(\delta_\pi, \theta) d\pi(\theta) \quad \textit{for all priors } \pi \textit{ on } \Theta.$$

The following lemma can be helpful for obtaining minimax estimators of a vector when minimax extimators of its components are known. The proof is obvious and omitted.

Lemma 4.3 *Let $X_i \sim^{ind} F_i(x; \theta_i)$, $i = 1, \ldots, k$ and let $\Theta = \{\theta \mid m_{i,1} \leq \theta_i, \leq m_{i,2}, i = 1, \ldots, k\}$ for known constants $(m_{i,1}, m_{i,2})$, $-\infty \leq m_{i,1} < m_{i,2} \leq \infty$, $i = 1, \ldots, k$. Then, if for $i = 1, \ldots, k$, $\pi_{n,i}$, $n = 1, 2, \ldots$, is a least-favourable sequence of piors for estimating $\theta_i \in [m_{i,1}, m_{i,2}]$ based on X_i with a loss function $L_i(d, \theta_i)$, $\prod_{i=1}^{k} \pi_{n,i}$, $n = 1, 2, \ldots$, is a least-favourable sequence of priors for estimating $\theta = (\theta_1, \ldots, \theta_k) \in \prod_{i=1}^{k} [m_{i,1}, m_{i,2}]$ based on $(X_1, \ldots, X_k)$ with the loss function $\sum_{i=1}^{k} L_i(d_i, \theta_i)$.*

A further useful result on minimax estimation can be found in Blumenthal and Cohen (1968b). Before stating their results, more needs to be said about the Pitman (1939) estimator of location mentioned in Chapter 2. This is an estimator of the parameter $\theta \in R^1$ based on a sample $X_1, \ldots, X_n$ from a distribution with Lebesgue density $f(x - \theta)$. When it has a finite risk function, it is the minimum-risk-equivariant estimator of θ for squared-error loss, as well as the Bayes estimator for squared-error loss with respect to the uniform distribution on $(-\infty, \infty)$. An explicit expression for the estimator is (see Pitman, 1939; Lehmann, 1983, p. 160; Lehmann and Casella, 1998, p. 154; or Berger, 1985, p. 400)

$$\delta_P(X_1, \ldots, X_n) = \frac{\int_{-\infty}^{\infty} u \prod_{i=1}^{n} f(x_i - u) du}{\int_{-\infty}^{\infty} \prod_{i=1}^{n} f(x_i - u) du}. \tag{4.2}$$

For the case of estimating a vector θ of location parameters based on independent samples $X_{i,1}, \ldots, X_{i,n}$, $i = 1, \ldots, k$, from distributions with Lebesgue densities $f(x - \theta_i)$, define the Pitman estimator δ_P of $\theta = (\theta_1, \ldots, \theta_k)$ as the Bayes estimator of θ for squared error loss with respect to the uniform prior on R^k. Then the i-th element, $i = 1, \ldots, k$, of the vector δ_P is the Pitman estimator of θ_i based on the i-th sample. Further, for any $k \geq 1$ and squared-error loss, the Pitman estimator of θ is minimax if its components have finite risk functions. A proof of this minimaxity for $k = 1$ can be found in Lehmann (1983, pp. 282–284) and in Lehmann and Casella (1998, pp. 340–342). These authors use a sequence of priors with uniform densities on $(-n, n)$, $n = 1, 2, \ldots$, and show it to be a least-favourable sequence of priors for estimating θ. They then use Lemma 4.2 to prove minimaxity. The minimaxity of the Pitman estimator when $k > 1$ follows from Lemma 4.3.

We are now ready to state the Blumenthal–Cohen results.

Lemma 4.4 *For $k \geq 1$ independent samples of equal size from distributions with Lebesgue densities $f(x - \theta_i)$, $i = 1, \ldots, k$, consider the estimation of $\theta = (\theta_1, \ldots, \theta_k) \in \Theta \subset R^k$ with squared-error loss. Let Θ be such that there exists a sequence of k-tuples of numbers $\{a_{n,1}, \ldots, a_{n,k}\}$, $n = 1, 2, \ldots$ for which*

$$\liminf_{n \to \infty} \{\theta \mid (\theta_1 + a_{n,1}, \ldots, \theta_k + a_{n,k}) \in \Theta\} = R^k. \tag{4.3}$$

Let δ_o be an estimator of θ for $\theta \in R^k$ and satisfying

$$R(\delta_o, \theta) \leq M \quad \text{for all } \theta \in \Theta, \tag{4.4}$$

where M is the constant risk (assumed to be finite) of the Pitman estimator of θ for $\theta \in R^k$. Then

$$\sup_{\theta \in \Theta} R(\delta_o, \theta) = M.$$

What this result tells us is that, under the stated conditions, $M(\mathcal{D}_o, \Theta) = M(\mathcal{D}_o, \Theta_o)$. This implies (by (2.17)) that $M(\mathcal{D}, \Theta) = M(\mathcal{D}_o, \Theta_o)$, i.e., the minimax values for the restricted and the unrestricted estimation of θ are equal and equal to the risk of the unrestricted Pitman estimator.

Blumenthal and Cohen (1968b) state and prove Lemma 4.4 for the special case where $k = 2$ but remark that a generalization to k dimensions is obvious – and it is. Kumar and Sharma (1988), apparently not having seen this Blumenthal–Cohen remark, state the generalization as a theorem.

A result similar to the Blumenthal-Cohen result holds for scale parameter estimation when $X_{i,j}$, $j = 1, \ldots, n_i$, $i = 1, \ldots, k$ are independent and the $X_{i,j}, j = 1, \ldots, n_i$ have Lebesgue density $f(x/\theta_i)/\theta_i, i = 1, \ldots, k$. Using scale-invariant squared-error loss, the Pitman estimator of $\theta = (\theta_1, \ldots, \theta_k) \in R_+^k$ is, when it has a finite risk function, the minimum-risk-equivariant estimator and also the Bayes estimator for the uniform prior for $\log \theta$ on $(-\infty, \infty)^k$. The i-th element, $i = 1, \ldots, k$, of this Pitman estimator is the Pitman estimator of θ_i based on $X_{i,1}, \ldots, X_{i,n_i}$ which is given by (see, e.g., Lehmann, 1983, p. 177; or Lehmann and Casella, 1998, p. 170)

$$\delta_i(X_{1,i}, \ldots, X_{i,n_i}) = \frac{\int_0^\infty t^{n_i} \prod_{j=1}^{n_i} f(tX_{i,j}) dt}{\int_0^\infty t^{n_i+1} \prod_{j=1}^{n_i} f(tX_{i,j}) dt}. \tag{4.5}$$

This estimator, being scale-equivariant, has a constant risk function. It is minimax for estimating $\theta \in R_+^k$ with scale-invariant squared-error loss. By Lemma 4.3 it is sufficient to show that this minimaxity holds for $k = 1$. And for this case the method of proof used by Lehmann (1983, pp. 282–284) and by Lehmann and Casella (1998, pp. 340–342) for the minimaxity in location problems with squared-error loss can easily be adapted to the scale-estimation case. The only condition is that the estimator has a finite risk function.

The analogue for scale estimation of the Blumenthal–Cohen result for location estimation is contained in the following lemma.

Lemma 4.5 *For $k \geq 1$ independent samples of equal size from distributions with Lebesgue densities $f(x/\theta_i)/\theta_i$, $\theta_i > 0$, $i = 1, \ldots, k$, consider the estimation of $\theta = (\theta_1, \ldots, \theta_k)$ with scale-invariant squared-error loss when θ is restricted to $\Theta \subset R_+^k$. Assume Θ is such that there exists a sequence of k-tuples of positive numbers $\{a_{1,n}, \ldots, a_{k,n}\}$ with*

$$\liminf_{n \to \infty} \{\theta \in R_+^k \mid (a_{1,n}\theta_1 \ldots, a_{k,n}\theta_k) \in \Theta\} = R_+^k.$$

Then, if δ_o is an estimator of θ for $\theta \in R_+^k$ and satisfying

$$R(\delta_o, \theta) \leq M \quad \text{for all } \theta \in \Theta, \tag{4.6}$$

where M is the constant risk of the Pitman estimator (assumed to be finite) for estimating $\theta \in R_+^k$, then

$$\sup_{\theta \in \Theta} R(\delta_o, \theta) = M.$$

Proof. We have not been able to find this result in the literature, but it follows easily from the scale-equivariance and minimaxity of the Pitman estimator and the techniques used by Blumenthal and Cohen (1968b) in their proof for the location case.   ♡

As in the location case, this result proves that, under the stated conditions, $M(\mathcal{D}, \Theta) = M(\mathcal{D}_o, \Theta_o)$, i.e., the minimax values for the restricted and unrestricted cases are equal and equal to the risk of the unrestricted Pitman estimator.

If the estimators δ_o of Lemma 4.4 satisfy (2.3), they are minimax for the restricted problem. In case they do not satisfy (2.3), they can (by the essential completeness of $\mathcal{D}$ in $\mathcal{D}_o$ with respect to Θ) be dominated on Θ by an estimator which does satisfy it. Such a dominator is then minimax for the restricted problem. The same remark holds for the estimator δ_o of Lemma 4.5. This proves the following corollary to the lemmas 4.4 and 4.5.

Corollary 4.1 *Under the conditions of Lemma 4.4, as well as under the conditions of Lemma 4.5, there exists, by the essential completeness of $\mathcal{D}$ in $\mathcal{D}_o$ with respect to Θ, a minimax estimator for estimating θ for $\theta \in \Theta$.*

Another useful result can be found in Hartigan (2004). For $X \sim \mathcal{N}_k(\theta, I)$, Θ a closed, convex proper subset of R^k with a non-empty interior and squared-error loss, he shows that the Pitman estimator δ_P of θ satisfies $R(\delta_P, \theta) \leq k$ with equality if and only if θ is an apex of Θ, i.e., any $\theta \in \Theta$ with the property that all tangent hyperplanes to Θ contain it. This result implies, e.g., that for estimating θ when $k = 1$ and $\Theta = [-m, m]$, the minimax value for squared-error loss, is less than 1.

Another set of useful results for solving problems of (admissible) minimax estimation consists of using lower bounds for the risk function of the problem. We mention some general results here. An example of how such bounds can be used in particular cases is described in Section 4.3.

A well-known lower bound for the risk function of an estimator δ can, when $\theta \in R^1$ and $h(\theta)$ is to be estimated, be obtained from the information inequality which says that, under regularity conditions (see, e.g., Lehmann, 1983, p. 122; or Lehmann and Casella, 1998, pp. 120–123),

$$\mathcal{E}_\theta(\delta(X) - h(\theta))^2 \geq b_\delta^2(\theta) + \frac{(h'(\theta) + b_\delta'(\theta))^2}{I(\theta)}, \tag{4.7}$$

where $b_\delta(\theta) = \mathcal{E}_\theta(\delta(X) - h(\theta))$ is the bias of the estimator δ of $h(\theta)$, $I(\theta)$ is the Fisher information

$$I(\theta) = \mathcal{E}_\theta \left(\frac{\partial}{\partial \theta} \log p_\theta(X) \right)^2$$

and the primes denote derivatives with respect to θ.

For the case where $\Theta = \{\theta \mid m \leq \theta < \infty\}$, Gajek and Kałuszka (1995, Section 3) present a class of lower bounds for the risk function of estimators of $h(\theta)$ with loss function $L(d, \theta) = (d - h(\theta))^2 w(\theta)$, where $0 < w(\theta) < \infty$ for all $\theta \in \Theta$. The estimator is based on $X \in R$ with Lebesgue density $f_\theta(x)$ and their estimators are not necessarily restricted to $\mathcal{H} = \{h(\theta) \mid \theta \in \Theta\}$. They suppose h to be a diffeomorphism and show that, for any function $H(x, \theta)$ and under the condition that $H(x, \theta)f_\theta(x)$ is, almost everywhere ν, absolutely continuous in θ on finite intervals (the authors ask only for continuously differentiable which is not enough for their proof to work)

$$\left. \begin{array}{l} \sup_{\theta \geq m} \mathcal{E}_\theta(\delta(X) - h(\theta))^2 w(\theta) \\[2em] \geq \dfrac{\left[\lim_{\theta \to \infty} h'(\theta) \int_{-\infty}^{\infty} H(x, \theta) f_\theta(x) dx \right]^2}{\lim_{\theta \to \infty} \{w(\theta)\}^{-1} \int_{-\infty}^{\infty} \dfrac{\left[\frac{\partial}{\partial \theta} (H(\theta, x) f_\theta(x)) \right]^2}{f_\theta(x)} dx}, \end{array} \right\} \tag{4.8}$$

provided the limits in the right-hand side exist.

Note that this bound does not depend on the estimator and is independent of m.

The special case where $H(x, \theta)$ is independent of x gives

$$\sup_{\theta \geq m} \mathcal{E}_\theta (\delta(X) - h(\theta))^2 w(\theta) \geq \lim_{\theta \to \infty} w(\theta) \frac{(h'(\theta)g(\theta))^2}{I(\theta)g^2(\theta) + (g'(\theta))^2}, \qquad (4.9)$$

where $g(\theta) = H(x, \theta)$. This inequality generalizes the following one which was proved by Gajek (1987)

$$\sup_{\theta \geq m} \mathcal{E}_\theta (\delta(X) - h(\theta))^2 w(\theta) \geq \lim_{\theta \to \infty} (h(\theta))^2 w(\theta) \frac{(h'(\theta))^2}{I(\theta)h^2(\theta) + ((h'(\theta))^2}, \qquad (4.10)$$

obtained by taking $g(\theta) = h(\theta)$.

For the case where $h(\theta) = \theta$ and $w(\theta) = 1$ for all $\theta \in \Theta$, (4.10) gives

$$\sup_{\theta \geq m} \mathcal{E}_\theta (\delta(X) - \theta)^2 \geq \lim_{\theta \to \infty} \frac{1}{I(\theta) + \theta^{-2}} = \lim_{\theta \to \infty} \frac{1}{I(\theta)},$$

a result which can also be found in Sato and Akahira (1995).

Gajek and Kałuszka (1995) say that, when $f_\theta(x) = f(x - \theta)$ as well as when $f_\theta(x) = f(x/\theta)/\theta$, their bounds are attainable. They prove this result for the case where $h(\theta) = \theta$ with $w(\theta) = 1$ for the location case and $w(\theta) = \theta^{-2}$ for the scale case. For the location case this gives (see their page 118)

$$\sup_{\theta \geq m} \mathcal{E}_\theta (\delta(X) - \theta)^2 \geq \int_{-\infty}^{\infty} (x - \bar{\theta})^2 f(x) dx, \qquad (4.11)$$

where $\bar{\theta} = \int_{-\infty}^{\infty} x f(x) dx$. Their conditions reduce in this case to existence of the second moment of X and the absolute continuity of $f(x)$. For the scale case their inequality becomes (see their page 121)

$$\sup_{\theta \geq m} \mathcal{E}_\theta \left(\frac{\delta(X)}{\theta} - 1 \right)^2 \geq 1 - \frac{\left(\int_0^\infty x f(x) dx \right)^2}{\int_0^\infty x^2 f(x) dx}, \qquad (4.12)$$

where their conditions reduce to existence of

$$\int_0^\infty \frac{\left(\int_0^y z(z a_1/a_2 - 1) f(z) dz \right)^2}{y^4 f(y)} dy$$

and absolute continuity of $x f(x)$, where $a_s = \int_0^\infty x^s f(x) dx$. These lower bounds are attained by, respectively,

$$\delta_L(x) = x \quad \text{and} \quad \delta_S(x) = x \frac{\int_0^\infty y f(y) dy}{\int_0^\infty y^2 f(y) dy}.$$

But, δ_L and δ_S are the Pitman estimators for estimating location and scale, respectively, when the parameter space is not restricted (see (4.2) and (4.5)). Note also that the Θ's satisfy the conditions of Lemma 4.4 and Lemma 4.5, respectively, implying that the restricted and unrestricted minimax values are equal. So, the inequalities (4.11) and (4.12) can also be obtained from these lemmas, but for these lemmas the only condition needed is the finiteness of the risk function of the Pitman estimator.

There are also lower bounds B_π, say, for the Bayes risk for estimating $h(\theta)$ with respect to a prior π on Θ. Such bounds imply that the minimax value for the problem is at least B_π for all π for which the bound holds. One lower-bound for the Bayes risk is the van Trees inequality (van Trees, 1968; see also, e.g., Gill and Levit, 1995 and Ruymgaart, 1996) which says that, under regularity conditions,

$$\mathcal{E}(\delta(X) - h(\theta))^2 \geq \frac{(\mathcal{E} h'(\theta))^2}{\mathcal{E} I(\theta) + I(\pi)}, \tag{4.13}$$

where $\mathcal{E}$ stands for expectation with respect to the joint distribution of X and θ and $I(\pi) = \mathcal{E}\left(\pi'(\theta)/\pi(\theta)\right)$, with $\pi(\theta)$ the Lebesgue density of the prior. Other lower bounds for the Bayes risk can, e.g., be found in Borovkov and Sakhanienko (1980), in Brown and Gajek (1990), in Vidakovic and DasGupta (1995) and in Sato and Akahira (1996). Sato and Akahira (1995) use the Borovkov and Sakhanienko (1980) and the Brown and Gajek (1990) results to obtain lower bounds for the minimax value.

Further, Hodges and Lehmann (1951) use the information inequality to prove (for the special case where $h(\theta) = \theta$ for all θ) the following result, where $CRB_\delta(\theta)$ denotes the right-hand side of (4.7).

Lemma 4.6 *For squared-error loss, if δ_o is an estimator of θ with equality in (4.7), if (4.7) holds for all estimators δ, then, if*

$$\{ CRB_\delta(\theta) \leq CRB_{\delta_o}(\theta) \quad \text{for all } \theta \in \Theta \} \Longrightarrow \{ b_\delta(\theta) \equiv b_{\delta_o}(\theta) \}$$

the estimator δ_o is admissible. If, moreover, δ_o has a constant risk function then it is minimax.

In restricted parameter spaces, minimax estimators do, typically, not have a constant risk function. So, Lemma 4.6 does not really help if one wants to check the minimaxity of a given estimator δ_o. However, in cases where $M(\mathcal{D}_o, \Theta_o) = M(\mathcal{D}, \Theta)$ and δ_o is an estimator satisfying (2.5) with a constant risk function on Θ_o, Lemma 4.6 can help find the minimax value. The same kind of comment applies to the corollary to Lemma 4.1 which implies that, if a Bayes estimator has a constant risk function, then it is minimax.

A result which can be useful for proving the admissibility of a(n) (minimax) estimator is Blyth's (1951) method, which can be formulated as follows (see, e.g., Lehmann, 1983, pp. 265–266; Lehmann and Casella, 1998, p. 380; or Berger, 1985, pp. 547–548).

Lemma 4.7 *Suppose that, for each estimator δ, the risk function $R(\delta, \theta)$ is continuous in θ for $\theta \in \Theta$. Then an estimator δ_o is admissible if there exists a sequence π_n of (possibly improper) measures on Θ such that*

a) $r_{\pi_n}(\delta_o) < \infty$ *for all n;*
b) *for any non-empty convex set $\Theta^* \subset \Theta$, there exists a constant $C > 0$ and an integer N such that*

$$\int_{\Theta^*} d\pi_n(\theta) \geq C \quad \text{for all } n \geq N;$$

c) $r_{\pi_n}(\delta_o) - r_{\pi_n}(\delta_n) \to 0$ *as $n \to \infty$.*

Finally, note that if δ_o is inadmissible minimax, then all of its dominators are minimax.

4.2 Minimax results when Θ is bounded

Minimax problems in restricted parameter spaces are difficult to solve – even more difficult than in unrestricted spaces and particularly when Θ is bounded. One reason for this is that, as already mentioned in Section 4.1, mimimax estimators in restricted parameter spaces, typically, do not have a constant risk function. So, looking for a Bayes estimator with a constant risk function very seldom helps. Further, for bounded Θ (restricted or not), there exists (see Wald (1950, Theorem 5.3)) a least favourable prior on Θ with finite support and a minimax estimator which is Bayes with respect to this prior. The number of points in this support increases with the "size" of Θ. The problem of finding these points and the prior probabilities can only seldom be solved analytically. As will be seen below, when $k = 1$ and $\theta = [m_1, m_2]$, analytical results have been obtained for small values of $m_2 - m_1$ or $(m_2/m_1) - 1$ for particular cases. For other particular cases, numerical results are available.

Concerning the result of Wald (1950, Theorem 5.3) quoted above, note that least favourable priors are not necessarily unique, i.e., a unique minimax estimator can be Bayes with respect to more than one least favourable prior. This will be shown in an example later in the present chapter, where a unique minimax estimator has an infinity of least favourable priors, some of which have finite support and some a Lesbegue density. So, Wald's result says that, under his conditions, at least one of the priors must have finite support.

The following result (see, e.g., Robert, 1997, pp. 60–61; Berger, 1985, p.353; or Kempthorne, 1987) sheds light on the relationship between constancy of the risk function of a Bayes estimator and the finiteness of the support of the priors.

Lemma 4.8 *Under the conditions of Lemma 4.1, if $\Theta \subset R^1$ is compact and the risk function of the minimax estimator δ_{π_o} is an analytic function of θ, either the least favourable prior π_o has finite support or the risk function of δ_{π_o} is constant.*

This lemma needs, in my opinion, to be reworded. Given that least favourable priors are not necessarily unique, "the least favourable prior" needs to be replaced by "all least favourable priors". Then the lemma implies that, if the conditions of Lemma 4.1 are satisfied, if $\Theta \subset R^1$ is compact and δ_{π_o} does have an analytic risk function which is not constant, then all least favourable priors have finite support.

The first minimax estimator for a bounded parameter space when $\Theta_o = (-\infty, \infty)$ was, it seems, obtained by Zinzius (1979, 1981) and Casella and Strawderman (1981). They consider (for squared-error loss) the case where $X \sim \mathcal{N}(\theta, 1)$ with $\theta \in [-m, m]$ for some known $m > 0$. They show that there exists an $m_o > 0$, such that, when $m \le m_o$,

$$\delta_m(X) = m \tanh(mX)$$

is a unique minimax estimator of θ and it is admissible.

They prove this result by taking a prior for θ with support $\{-m, m\}$ and choosing this prior in such a way that the corresponding Bayes estimator δ_m of θ satisfies $R(\delta_m, -m) = R(\delta_m, m)$. This prior puts equal mass on $\{-m\}$ and $\{m\}$. In order for δ_m to be minimax, and thus the prior to be least favourable, it is sufficient (see Lemma 4.1) that

$$R(\delta_m, \theta) \le R(\delta_m, m) \quad \text{for all } \theta \in [-m, m].$$

Casella and Strawderman first show that

$$\max_{\theta \in [-m,m]} R(\delta_m, \theta) = \max(R(\delta_m, 0), R(\delta_m, m)).$$

They then study the function $g(m) = R(\delta_m, 0) - R(\delta_m, m)$ and show, using Karlin's (1957) Theorem 3 and Corollary 2, that $g(m)$ changes sign only once when m moves from 0 to ∞. This change of sign is from negative to positive, which implies that there exists a unique $m_o > 0$ such that $g(m) \le 0$ when $m \le m_o$. And this proves their result. Numerically they find $m_o \approx 1.056742$.

They also show that, for $m > m_o$, the estimator δ_m is not minimax.

Zinzius studies the second derivative of the risk function of δ_m and finds an $m_o > 0$ such that a lower bound for this derivative is positive for all $\theta \in [-m, m]$ when $m \leq m_o$. This gives him, numerically, $2m_o \approx 1.20$. So, Zinzius's m_o is not best possible in the sense that there are m's larger than his m_o for which his Bayes estimator is minimax. Zinzius also has numerical results for $2m \leq 2.1$, i.e., for values of m less than the 1.056742 of Casella and Strawderman. An improvement on the Zinzius result was obtained by Eichenauer, Kirschgarth and Lehn (1988), but their result is weaker than the Casella–Strawderman one, i.e., also not best-possible.

These same two techniques were later used to solve many other mimimax problems for the case where Θ is a "small" closed convex subset of R^k. The minimax estimators are Bayes with respect to a prior for θ whose support is contained in the boundary $B(\Theta)$ of Θ. It is chosen in such a way that this Bayes estimator, δ_m, has a risk function which is constant on $B(\Theta)$. Then it is shown that, for "small enough" Θ, the risk function of δ_m attains its maximum value on Θ for (a) value(s) of $\theta \in B(\Theta)$. In most cases the maximum "size" of Θ can only be obtained by numerical methods. The techniques used are, in most cases, the one used by Zinzius and in those cases the m_o obtained is not best possible.

We will not discuss these results in detail, but list them and make comments on some of them.

1) The bounded normal mean problem where $-m \leq \theta \leq m$ for a given $m > 0$ was generalized by using the loss function $|d - \theta|^p$ by Bischoff and Fieger (1992) for $p \geq 2$ and by Eichenauer-Herrmann and Ickstadt (1992) for $p > 1$. For the special case where $p = 2$, their result is weaker than the Casella–Strawderman one, but an improvement over the one of Zinzius. This can be seen from the results of computations of their $m_o(p)$ (results which can also be found in Bischoff and Fieger, 1992). Eichenauer-Herrmann and Ickstadt (1992) study the problem of finding the minimax estimator also numerically and find, up to numerical accuracy, the best possible $m_o(p)$. For $p = 2$ these numerical results give the (best-possible) result of Casella–Strawderman.

Extensions to general location problems where X has Lebesgue density $f(x - \theta)$ with the loss function $|\theta - d|^p$, $p > 1$ can be found in Eichenauer-Herrmann and Ickstadt (1992). They also show there that, when $p = 1$, no two-point prior can be least favourable for this problem.

Results for the normal mean problem with $X \sim \mathcal{N}(\theta, \sigma^2)$, σ^2 known, $\theta \in [-m, m]$, and the linex loss function $e^{a(d-\theta)/\sigma} - a(d-\theta)/\sigma - 1$, where $a \neq 0$

is a known constant, can be found in Bischoff, Fieger and Wulfert (1995). They show that, when $0 < m/\sigma \leq m_o = \min\{a(\sqrt{3}+1)/2, \log 3/(2a)\}$, there exists a unique two-point prior with mass π on $\{-m\}$ and $1-\pi$ on $\{m\}$ for which the Bayes estimator is minimax. This estimator is given by

$$\delta_m(X) = \frac{\sigma}{a} \log \frac{g_m(X)}{g_m(X-a\sigma)},$$

where $g_m(x) = \pi e^{-mx/\sigma^2} + (1-\pi)e^{mx/\sigma^2}$. That this m_o is not best-possible can be seen from their proof as well as from their numerical results where they give (among other things) the maximum value of m for which δ_m is minimax. For $a = .5$ and $\sigma = 1$, e.g., this maximum value is 1.0664, while $m_o = .1830$. (There is a misprint in the author's formula for m_o. They give $m_o = \min\{a(\sqrt{3}-1)/2, \log 3/(2a)\}$);

2) The above-mentioned result of Bischoff, Fieger and Wulfert (1995) has been extended to the general case of a sample $X_1, \ldots, X_n$ from a distribution P_θ, $\theta \in [m_1, m_2]$, $m_1 < m_2$, by Wan, Zou and Lee (2000). They apply their results to minimax estimation of a Poisson mean θ when $\theta \in [0, m]$ and $L(d, \theta) = e^{a(d-\theta)} - a(d-\theta) - 1$ for a known constant $a \neq 0$. They give an $m_o > 0$ such that, when $0 < m < m_o$, there exists a two-point prior with mass π on $\{0\}$ and mass $1-\pi$ on $\{m\}$ which is least favourable. The Bayes estimator with respect to this prior is then a minimax estimator. It is given by

$$\delta(X_1, \ldots, X_n) = \begin{cases} \frac{1}{a} \log \dfrac{\pi + (1-\pi)e^{-nm}}{\pi + (1-\pi)e^{-(n+a)m}} & \text{when } \sum_{i=1}^n X_i = 0 \\ m & \text{when } \sum_{i=1}^n X_i > 0. \end{cases}$$

This result is not best-possible;

3) The exponential location problem where X has density $e^{-(x-\theta)}$, $x > \theta$ and θ is to be estimated with squared-error loss was solved by Eichenauer (1986). For the parameter interval $[0, m]$, he proves that $m \leq m_o$, where $m_o \approx .913$, is necessary and sufficient for the two-point prior with

$$\text{mass } \frac{1}{1 + e^{-m/2}} \text{ on } \{0\} \text{ and mass } \frac{e^{-m/2}}{1 + e^{-m/2}} \text{ on } \{m\}$$

to be least favourable. The minimax estimator is then given by

$$\delta_m(X) = \frac{m}{1 + e^{-m/2}} I(X \geq m).$$

As Eichenauer notes, his result can be extended to the case where a sample $X_1, \ldots, X_n$ is available. In that case $X = \min(X_1, \ldots, X_n)$ is sufficient for θ and has density $ne^{-n(x-\theta)}I(x > \theta)$. This Eichenauer (1986) result can also be found in Berry (1993);

4) For a sample from a distribution with a Lebesgue density with support
   $[\theta, \theta+1]$, $\theta \in [0, m]$, and with a convex loss function, Eichenauer-Herrmann
   and Fieger (1992) give sufficient conditions on the loss function, on m and
   on the two-point prior on $\{0, m\}$ to be least favourable. As an example of
   their results they take $L(d, \theta) = |d - \theta|$ and $X_1, \ldots, X_n$ a sample form a
   uniform distribution on $[\theta, \theta + 1]$. They show that, for each $n \geq 1$, there
   exists an $m_n^* \in (0, 1)$ such that, for $0 < m < m_n^*$, the Bayes estimator with
   respect to the uniform prior on $\{0, m\}$ is minimax. This estimator is given
   by

$$\delta_m(X) = \begin{cases} 0 & \text{when} & X \in [0, 1]^n \backslash [m, 1]^n \\ m & \text{when} & X \in [m, m + 1]^n \backslash [m, 1]^n \\ m/2 & \text{otherwise,} \end{cases}$$

   where $X = (X_1, \ldots, X_n)$. They also provide a table giving the values of
   m_n^* for several values of n. As a second example they mention the case
   where the X_i have density $(e - 1)e^{\theta+1-x} I(\theta \leq x \leq \theta + 1)$;

5) For a sample $X_1, \ldots, X_n$ from a uniform distribution on the interval
   $[-\alpha\,\theta^\gamma, \beta\,\theta^\gamma]$, where α, β and γ are known with $\alpha, \beta \geq 0$ and $\alpha+\beta, \gamma > 0$,
   Chen and Eichenauer (1988) use squared-error loss for estimating θ when it
   is restricted to the interval $[c, cm]$ with $c > 0$ and $m > 1$. They show that,
   given α, β, γ, c, and n, there exists an $m^* > 1$ such that, for $1 < m < m^*$,
   the Bayes estimator of θ with respect to the two-point prior with

$$\text{mass} \; \frac{1}{1 + m^{\gamma n/2}} \; \text{on} \; \{c\} \; \text{and mass} \; \frac{m^{\gamma n/2}}{1 + m^{\gamma n/2}} \; \text{on} \; \{cm\}$$

   is minimax. This estimator is given by

$$\delta_m(X) = \begin{cases} c\,\dfrac{m^{\gamma n/2} + m}{m^{\gamma n/2} + 1} & \text{when } X \in [-\alpha c^\gamma, \beta c^\gamma]^n \\ cm & \text{otherwise,} \end{cases}$$

   where $X = (X_1, \ldots, X_n)$;

6) Several papers study the estimation of $h(\theta)$ when X has Lebesgue density
   $f(x/\theta)/\theta$ with θ restricted to the interval $[c, cm]$, for $c > 0$ and $m > 1$.
   Eichenauer-Herrmann and Fieger (1989) use squared-error loss and sup-
   pose h to be twice continuously differentiable with $h'(\theta) \neq 0$ for $\theta \geq c$.
   Bischoff (1992) uses the loss function $|d - h(\theta)|^p$ with $p \geq 2$. He supposes
   h to be strictly monotone. For estimating θ, van Eeden and Zidek (1999)
   use scale-invariant squared-error loss. Each of these papers gives sufficient
   conditions for the Bayes estimator with respect to a prior on the boundary
   of Θ to be minimax. In van Eeden and Zidek (1994b) the (at that date
   still unpublished) results of van Eeden and Zidek (1999) are used for the
   special case of the F distribution;

7) Several results have been obtained for the case where X is $\mathcal{N}_k(\theta, I)$. Berry (1990) takes Θ to be a sphere or a rectangle and uses squared-error loss to estimate θ. For the case of a rectangle he remarks, as is also remarked above, that the minimax estimator of the vector is the vector of minimax estimators of the components. So for this case the results of Zinzius (1979, 1981) and of Casella and Strawderman (1981) can be used to obtain minimax estimators of the vector parameter when the rectangle is "small". For the case of a sphere of radius m, Berry (1990) shows that there exists an $m_o(k) > 0$ such that, when $0 < m \leq m_o(k)$, the Bayes estimator δ_m with respect to the uniform distribution on the boundary of Θ is minimax. For $k \in \{2, 3\}$, he obtains explicit expressions for the minimax estimator. Numerically he finds $m_o(2) \approx 1.53499$ and $m_o(3) \approx 1.90799$. Berry's results are, like the ones of Casella and Strawderman, best-possible, i.e., his estimators are not minimax when $m > m_o(k)$. Marchand and Perron (2002) generalize the Berry (1990) results for the sphere to k ($k \geq 4$) dimensions and show that $m_o(k) \geq \sqrt{k}$. This result then implies that $m_o(k) \geq \sqrt{k}$ for all $k \geq 1$. Bischoff, Fieger and Ochtrop (1995) take $k = 2$, squared error loss and θ restricted to an equilateral triangle. Their result that for a small enough triangle there exists a unique minimax estimator of the vector θ is not best-possible;

8) In a very general setting, Bischoff and Fieger (1993) investigate whether, for absolute-error loss, there exists a two-point least favourable prior on the boundary of Θ. They give examples of existence as well as of non-existence of such priors;

9) Another example where minimax estimators for "small" parameter spaces have been obtained is the case where $X \sim \text{Bin}(n, \theta)$ with $\theta \in [m_1, m_2]$ with $0 \leq m_1 < m_2 < 1$. Marchand and MacGibbon (2000) use squared-error loss as well as the normalized loss function $(d - \theta)^2 / (\theta(1 - \theta))$ and give necessary and sufficient conditions on m_1 and m_2 for the Bayes estimator with respect to a prior on $\{m_1, m_2\}$ to be minimax. For normalized loss they also have results for "moderate" values of m;

10) DasGupta (1985) considers, in a very general setting, the estimation, under squared-error loss, of a vector $h(\theta)$ when θ is restricted to a small bounded convex subset Θ of R^k. He gives sufficient conditions under which the Bayes estimator with respect to the least favourable prior on the boundary of Θ is minimax for estimating $h(\theta)$. In one of his examples he gives minimax estimators of θ for the case where $X \sim U(\theta - 1/2, \theta + 1/2)$, $\theta \in [-m, m]$, $m \leq 1/4$ and for the case where $X \sim U(0, \theta)$, $\theta \in [m_1, m_2]$, $m_2/m_1 \leq 2/(\sqrt{5} - 1)$. Two other examples are $X \sim \mathcal{N}(\theta, 1)$ with $|\theta| \leq m$ and $X \sim \text{Bin}(n, \theta)$ with $|\theta - 1/2| \leq m$ for a known $m < 1/2$. The above-mentioned results on the existence of a minimax estimator for "small" Θ when the loss is squared error can be obtained from DasGupta's (1985) results and most authors do refer to him. Chen and Eichenauer (1988) note that DasGupta's result for the $\mathcal{U}(0, \theta)$ distribution is the special case of their result with $\alpha = 0$ and $\beta = \gamma = n = 1$. Further, the minimax

estimator obtained by Eichenauer-Herrmann and Fieger (1992, p. 34) for estimating a bounded θ with absolute-error loss when $X \sim \mathcal{U}(\theta, \theta + 1)$ is, for $n = 1$, the same as the minimax estimator DasGupta obtains for the same X with squared-error loss and the maximum sizes of Θ for which their results hold are also the same. For the bounded-normal-mean case DasGupta finds that $m \leq .643$ is sufficient for the Bayes estimator with respect to a uniform prior on the boundary of Θ to be minimax. This result is slightly better than Zinzius's (1979, 1981) result. For his binomial example DasGupta finds that for any Θ no larger than $[.147, .853]$ the Bayes estimator with respect to a uniform prior on the boundary of Θ is minimax;

11) A result unifying many of the above ones can be found in Bader and Bischoff (2003). They have X with Lesbegue density $f(x; \theta)$, $\theta \in [m_1, m_2]$ and a loss function of the form $L(d, \theta) = l(d - \theta)$. They then give conditions on l and the densities for a two-point prior on $\{m_1, m_2\}$ to be least favourable.

Remark 4.1. Most of the above scale-parameter results, i.e., those in 5), 6) and 10), are derived for loss functions which are not scale-invariant. The only exception is the result of van Eeden and Zidek (1994b). They use scale-invariant squared-error loss.

Remark 4.2. It can easily be seen that, in each of the above problems where the support of the distribution of X depends on θ (i.e., those in 3) - 5) and the uniform cases of DasGupta), the estimator of θ satisfies (see Chapter 2) the "extra" restriction that the probability that X is in its estimated support equals 1 for each $\theta \in \Theta$.

For a binomial parameter restricted to the interval $[m, 1 - m]$, as well as for a Poisson mean restricted to the interval $[0, m]$, analytical as well as numerical results have been obtained for abitrary m.

For the case where X has a $\text{Bin}(n, \theta)$ distribution and $\theta \in [m, 1 - m]$, $0 < m < 1/2$, Moors (1985) gives the minimax estimator and all its least favourable priors for $n = 1, 2$ and 3 for squared-error as well as for normalized loss. For larger values of n, up to 16, he has numerical results. Moors also shows, for squared-error loss, that there exists a least favourable prior with $\leq [(n + 3)/4] + 1$ points in its support.

Berry (1989), using an approach different from the one of Moors, gives the minimax estimators and least favourable priors for squared-error loss for $n = 1, 2, \ldots, 6$. However, for $n = 5, 6$, Berry's results hold only for small m.

In the special case where $n = 1$ the results of Moors (1985) and Berry (1989) imply that, when $(2 - \sqrt{2})/4 \leq m < 1/2$, the estimator $\delta(0) = 2m(1 - m)$,

$\delta(1) = 1 - \delta(0)$ is minimax. This result was also obtained by DasGupta (1985). The complete results for $n = 1$ and $n = 2$ can also be found in Zou (1993). (This article by Zou is written in Chinese. My thanks to (Xiaogang) Steven Wang for the translation.)

An interesting result, mentioned by Moors and by Berry as well as by Zou, is that for small enough m, the unique mimimax estimator for the parameter space $[0, 1]$ is minimax for the parameter space $[m, 1 - m]$. A sufficient condition for this result to hold is (as Moors shows) that $(n + \sqrt{n}/2)(n + \sqrt{n}) \leq 1 - m$ and that there exists a prior on $[m, 1 - m]$ whose moments μ_i satisfy

$$\mu_{i+1} = \frac{i + \sqrt{n}/2}{i + \sqrt{n}} \mu_i \qquad i = 0, 1, \ldots, n.$$

Zubrzycki (1966) mentions that Dzo-i (1961) has this same result. It is an example of a case where a minimax estimator for a restricted parameter space has a constant risk function. It is also an example where the least favourable distribution is not unique, because, given that it is minimax for $\theta \in [0, 1]$, it is Bayes with respect to the Beta $((\sqrt{n})/2, (\sqrt{n})/2)$ prior on $[0, 1]$ and, given that it is minimax for $\theta \in [m, 1 - m]$, it is Bayes with respect to some prior on $[m, 1 - m]$, which, of course, is a prior on $[0, 1]$. For $n = 1$, e.g., Moors shows that, when $m \leq (2 - \sqrt{2})/4$, there exist two symmetric discrete priors on $[m, 1 - m]$, each giving the same Bayes estimator which is minimax for the parameter space $[m, 1 - m]$ as well as for the parameter space $[0, 1]$. These priors are given by

$$\pi_1 \text{ with mass } \frac{1}{2} \text{ on each of } \frac{2 \pm \sqrt{2}}{4}$$

and

$$\pi_2 \text{ with mass } \frac{1}{4(2m - 1)^2} \text{ on each of } m \text{ and } 1 - m$$

and the rest of the mass on $1/2$.

The Bayes estimator takes the values $1/4$ for $x = 0$ and $3/4$ for $x = 1$ and this estimator is also Bayes with respect to the Beta $(1/2, 1/2)$ prior. This example with the prior π_1 can also be found in Berry (1989).

The fact that, for the binomial case with squared-error loss, the least favourable prior is not necessarily unique is not surprising given that (see Moors, 1985, Section 5.5) the Bayes risk for a given prior is determined by the first n of its moments.

For X a random variable with a Poisson distribution with mean $\theta \in [0, m]$, Johnstone and MacGibbon (1992) use the loss function $(d - \theta)^2/\theta$ and give minimax estimators and least favourable priors for $0 < m \le m_o$, where $m_o \approx 1.27$. For $m \le m_1$, where $m_1 \approx .57$ the least favourable prior has $\{0\}$ as its support and for $m_1 < m \le m_o$ the prior is a two-point prior with support $\{0, m\}$. They use numerical methods to obtain the minimax estimator and least favourable priors for selected values of m between .100 and 11.5. As already mentioned above, Wan, Zou and Lee (2000) consider this Poisson problem with the linex loss function.

For $X \sim \mathcal{N}(\theta, 1)$ with $|\theta| \le m$ for a known positive m, Zeytinoglu and Mintz (1984) obtain an admissible minimax estimator δ_m of θ for the loss function

$$L(d, \theta) = I(|d - \theta| > e), \tag{4.14}$$

where $e > 0$ is known and $m > e$. For the case where $e < m \le 2e$, e.g., they show the estimator

$$\delta_m(X) = \begin{cases} -(m - e) & \text{when } X \le -(m - e) \\ X & \text{when } -(m - e) < X < m - e \\ m - e & \text{when } X \ge m - e \end{cases}$$

to be admissible minimax for estimating θ when θ is restricted to the interval $[-m, m]$. It is Bayes with respect to the (least favourable) prior $I(\theta \in [-m, m - 2e] \cup [2e - m, m])(m - e)/4$. It is also the MLE of θ when θ is restricted to the interval $[-(m - e), m - e]$.

Finally, some remarks on a paper by Towhidi and Behboodian (2002). They claim to have a minimax estimator of a bounded normal mean under the so-called reflected-normal loss function which is given by

$$L(d, \theta) = 1 - e^{-(d - \theta)^2/(2\gamma^2)},$$

where γ is known and $-m \le \theta \le m$. They start with a prior on $\{-m, m\}$ and find that, when $\gamma > 2m$, the Bayes estimator δ of θ based on $X \sim \mathcal{N}(\theta, 1)$ is the solution to the equation

$$(m - \delta(x))p(x)e^{2m\delta(x)/\gamma^2} = (m + \delta(x))(1 - p(x)), \tag{4.15}$$

where $p(x) = P(\theta = m \mid X = x)$. However, they do not say anything about the number of roots to (4.15). This problem with their results can, however, be solved. It is not difficult to see that (4.15) has exactly one solution and that this solution is the Bayes estimator when $\gamma > 2m$. A much more serious problem with this paper of Towhidi and Behboodian is that their proof that this Bayes estimator is minimax is incorrect. In fact, they incorrectly apply Karlin's (1957) theorem on the relationship between the number of changes of sign of an integral and the number of changes of sign of its integrand. I do not know how to correct this mistake and extensive correspondence with the authors about this problem has not resolved it. The proofs of their theorems concerning the analogous problem of minimax estimation of a bounded scale parameter contain the same errors.

Remark 4.3. Kałuszka (1986) has results for minimax estimation, with scale-invariant squared-error loss, of the scale parameter θ of a gamma distribution when $\theta \in (0, m]$ for a known positive m. These results are presented in Section 4.3 together with his results for the case where $\theta \geq m$ for a known positive m.

4.3 Minimax results when $k = 1$ and Θ is not bounded

In this section we look at the problem of minimax estimation of θ when $\Theta_o = [m', \infty)$ and $\Theta = [m, \infty)$ with $-\infty \leq m' < m$.

There are few results for this case and, among these, many only give estimators for the $(\mathcal{D}_o, \Theta)$ problem. Of course, as mentioned before, such estimators can (when $\mathcal{D}$ is complete in $\mathcal{D}_o$) be replaced by estimators satisfying (2.3) which dominate them on Θ. Such dominators are minimax for the $(\mathcal{D}, \Theta)$ problem.

Further, when the risk function of a $(\mathcal{D}_o, \Theta_o)$-minimax estimator is constant on Θ_o, this estimator is $(\mathcal{D}_o, \Theta)$-minimax, making the $(\mathcal{D}_o, \Theta)$-minimax problem trivial if the only purpose is to find a minimax estimator. However, a careful study of the minimax problems for $(\mathcal{D}_o, \Theta_o)$ and $(\mathcal{D}_o, \Theta)$, might still be useful because it might provide a class of $(\mathcal{D}_o, \Theta)$-minimax estimators some of which are $(\mathcal{D}, \Theta)$-minimax. As we will see below, for the case of a lower-bounded scale parameter of a gamma distribution, several authors have obtained $(\mathcal{D}, \Theta)$-minimax estimators this way, although none of them explicitly states which minimax problem they are solving, nor do they indicate which of their estimators satisfy (2.3).

Below known results and (outlines of) their proofs on (admissible) minimax estimation with Θ_o and Θ as given above are presented for the case where (i) θ is a normal mean; (ii) θ is a scale parameter of a gamma distribution, a transformed χ^2-distribution (as introduced by Rahman and Gupta, 1993), or an F-distribution; (iii) θ is a location parameter of a uniform distribution; (iv) θ is a scale parameter of a uniform distribution and (v) θ is an exponential location parameter. For this last case, results for the case where θ is upper-bounded are also presented. Finally, some new results for a lower-bounded mean of a Poisson distribution are given.

First, let $X \sim \mathcal{N}(\theta, 1)$ with $\theta \geq 0$ and squared-error loss. The earliest result for this case is by Katz (1961), who gives

$$\delta_K(X) = X + \frac{\phi(X)}{\Phi(X)}$$

as an admissible minimax estimator of θ, where ϕ and Φ are, respectively, the density and distribution function of $X - \theta$. This estimator is easily seen to be the Pitman estimator of θ under the restriction $\theta \geq 0$, i.e. the Bayes estimator of θ for the uniform prior on $[0, \infty)$. For a proof of the minimaxity of his estimator Katz uses Lemma 4.2 with

$$\lambda_n(\theta) = \frac{1}{n}e^{-\theta/n}I(\theta \geq 0) \qquad \theta \geq 0, n = 1, 2, \ldots,$$

as the sequence of prior densities for θ. He then needs to prove that $R(\delta_K, \theta) \leq r$ for all $\theta \geq 0$, where r is the limit of the sequence of Bayes risks. He shows that $r = 1$ and what then needs to be shown is that

$$\mathcal{E}_\theta(\delta_K(X) - \theta)^2 \leq 1 \qquad \text{for all } \theta \geq 0. \tag{4.16}$$

His proof of (4.16) is not correct. However, simple, somewhat lenghty, algebra shows that

$$2\mathcal{E}_\theta(X - \theta)\frac{\phi(X)}{\Phi(X)} + \mathcal{E}_\theta\left(\frac{\phi(X)}{\Phi(X)}\right)^2 = -\theta\int_{-\infty}^{\infty}\frac{\phi(x)}{\Phi(x)}\phi(x - \theta)dx \leq 0 \text{ for } \theta \geq 0,$$

from which the result follows. Note that Katz's result also shows that $M(\mathcal{D}_o, \Theta_o) = M(\mathcal{D}, \Theta)$ and that their common value is 1.

There is however an easier way to prove the minimaxity of δ_K. First note that X, the Pitman estimator of the unrestricted θ, has a risk function which is constant on Θ_o. So, by Lemma 4.4 it is sufficient to prove only (4.16) and no sequence of priors needs to be guessed at. A very simple proof of (4.16) can be obtained by using Kubokawa's (1994b) integral-expression-of-risk method. This method, applied to location problems, gives sufficient conditions for $\delta(X) = X + \psi(X)$ to dominate X. Assuming ψ to be absolutely continuous with $\psi(x) \to 0$ as $x \to \infty$, he writes

$$(x - \theta)^2 - (x - \theta + \psi(x))^2 = -\frac{1}{2}\int_x^{\infty}(x - \theta + \psi(t))\psi'(t)dt.$$

This gives

$$\mathcal{E}_\theta(X - \theta)^2 - \mathcal{E}_\theta(X - \theta + \psi(X))^2 = -\frac{1}{2}\int_{-\infty}^{\infty}\psi'(t)\int_{-\infty}^{t}(x - \theta + \psi(t))f(x - \theta)dx,$$

where $f(x - \theta)$ is the density of X. Then, assuming ψ to be nonincreasing, a sufficient condition for δ to dominate X on $[0, \infty)$ is that

$$\int_{-\infty}^{t-\theta}(x + \psi(t))f(x)dx \leq 0 \qquad \text{for all } t \text{ and all } \theta \geq 0,$$

which is equivalent to

$$\psi(t) \int_{-\infty}^{t-\theta} f(x)dx \geq - \int_{-\infty}^{t-\theta} xf(x)dx \qquad \text{for all } t \text{ and all } \theta \geq 0.$$

For our normal means case this condition becomes

$$\psi(t) \geq \frac{\phi(t)}{\Phi(t)} \qquad \text{for all } t,$$

which proves (4.16), because $\phi(t)/\Phi(t)$ satisfies the conditions imposed on ψ.

A possibly still easier way to show that δ_K is minimax is to use Lemma 4.4 and Hartigan (2004) as follows. Because the risk function of X (the unrestricted Pitman estimator) equals 1, Lemma 4.4 tells us that it is sufficient to show that $R(\delta_K, \theta) < 1$ for $\theta > 0$ and $= 1$ for $\theta = 0$. And that is exactly what Hartigan (2004) shows.

But, of course, neither the Blumenthal–Cohen, the Kubokawa, nor the Hartigan results were available to Katz.

Katz (1961) proves the admissibility of δ_K by using Blyth's (1951) method (Lemma 4.7).

Remark 4.4. We note here that Katz's (1961) paper also contains results on admissible minimax estimation of a lower-bounded expectation parameter for the more general case of an exponential-family distribution. However, he does not verify whether his Bayes estimators have finite Bayes risk and, in fact, in general they do not. So, for such cases, his proofs are incorrect. (See van Eeden, 1995, for a specific example.) As already mentioned in Remark 3.6, Chang (1982) also quotes and uses, without any comments, the possibly incorrect admissibility results of Katz (1961) for a lower-bounded gamma scale parameter, as well as for a lower-bounded mean of a Poisson distribution.

Still for the lower-bounded normal mean case with squared-error loss, the MLE of θ is the projection of X unto $\Theta = [0, \infty)$. So the MLE dominates X on Θ and the fact that $M(\mathcal{D}_o, \Theta_o) = M(\mathcal{D}, \Theta)$, then implies that the MLE is minimax. As already seen in Chapter 3, Section 3.1, the MLE for the lower-bounded normal mean is inadmissible for squared-error loss. Its minimaxity then implies that all its dominators, e.g., the ones obtained by Shao and Strawderman (1996b), are also minimax.

The next case we look at is the minimax estimation of a lower-bounded scale parameter of a gamma distribution. Let X have density

$$f_\theta(x) = \frac{1}{\theta^\alpha \Gamma(\alpha)} x^{\alpha-1} e^{-x/\theta} \qquad x > 0,$$

where $\alpha > 0$ is known and $\theta \geq m$ or $\theta \in (0, m]$ for a known $m > 0$. The loss function is scale-invariant squared-error loss, i.e., $L(d, \theta) = ((d/\theta) - 1)^2$.

Minimax estimators as well as admissible estimators have been obtained for this case by Kałuszka (1986, 1988). He estimates θ^ρ for a given $\rho \neq 0$ when $\theta \geq m$, as well as when $\theta \in (0, m]$, for a given $m > 0$. Going through his proofs, it can be seen that he solves the $(\mathcal{D}_o, \Theta)$-problem, i.e. his estimators do not satisfy (2.3) and he compares them on Θ. He first shows that, when $-\rho < \alpha/2$, $\rho \neq 0$,

$$V_\rho = 1 - \frac{\Gamma^2(\alpha + \rho)}{\Gamma(\alpha)\Gamma(\alpha + \rho)}$$

is an upper bound on the $(\mathcal{D}_o, \Theta)$-minimax values for these two problems. He obtains this result by using Lemma 4.2 with truncated generalized gamma densities for a sequence of priors for θ^ρ. Then, by obtaining estimators whose minimax risk equals V_ρ, he shows that V_ρ is the minimax value for each of the two problems. By the essential completeness of $\mathcal{D}$ in $\mathcal{D}_o$ with respect to Θ, V_ρ is also the minimax value for the $(\mathcal{D}, \Theta)$-problem.

His minimax estimators are given by

$$\delta(X) = \frac{\Gamma(\alpha + \rho)}{\Gamma(\alpha + 2\rho)} X^\rho + b,$$

where $b = 0$ when $0 < \theta \leq m$, $\rho > 0$ and also when $\theta \geq m$, $0 < -\rho < \alpha/2$. These estimators clearly do not satisfy (2.3). However, when $0 < \theta \leq m$, $0 < -\rho < \alpha/2$ and also when $\theta \geq m$, $\rho > 0$, the condition on b for δ to be minimax becomes

$$0 \leq b \leq 2m^\rho \left(1 - \frac{\Gamma^2(\alpha + \rho)}{\Gamma(\alpha)\Gamma(\alpha + 2\rho)}\right)$$

and then, if $\Gamma^2(\alpha + \rho)/(\Gamma(\alpha)\Gamma(\alpha + 2\rho)) \leq 1/2$, there exist b such that δ is minimax and is in $\mathcal{D}$. However this condition is rather restrictive. For $\rho = 1$, e.g., we are estimating θ and the condition on α becomes $2 < \alpha \leq 3$.

Kałuszka (1986) does not say anything about the admissibility of his minimax estimators other than to give conditions under which they are admissible (for squared-error loss) among estimators of the form $aX^\rho + b$.

Kałuszka (1986) also gives admissible estimators for each of his problems. These estimators satisfy (2.3). This is difficult to see from their formulas. It follows, however, directly from the fact that they are admissible, because (given that $\mathcal{D}$ is essentially complete in $\mathcal{D}_o$ with respect to Θ) estimators which do not satisfy (2.3) are inadmissible for the $(\mathcal{D}, \Theta)$ problem.

Kałuszka does not say whether his admissible estimators are minimax or not. Kałuszla's (1988) generalizes his results to the problem of minimax estimation of functions $h(\theta)$.

For the special cases where $\rho = 1$ and $\rho = -1$, Zubrzycki (1966) has Kałuszla's (1986) minimax results. Further, Ghosh and Singh (1970) take $\rho = -1$, $\alpha = 3$ and have results for $0 < \theta \leq m$ as well as for $\theta \geq m$. They solve the $(\mathcal{D}_o, \Theta)$ estimation problem with scale-invariant squared-error loss for these Θ's. This gives them the minimax value $M(\mathcal{D}, \Theta)$ $(=M(\mathcal{D}_o, \Theta))$. Their minimax estimator does not satisfy (2.3), but any dominator of it which is in $\mathcal{D}$ is a minimax estimator for the $(\mathcal{D}, \Theta)$ problem.

For the case where $\rho = 1$ and $\theta \geq m > 0$, van Eeden (1995) gives an admissible minimax estimator for scale-invariant squared-error loss. Her results were obtained independently of those of Kałuszka (1986, 1988), but her estimator is a special case of his admissible estimators, of which (as already said) he does not prove minimaxity. This admissible minimax estimator is given by

$$\delta_{vE}(X) = \frac{X}{\alpha + 1}\left(1 + \frac{g_{\alpha+2}(X/m)}{G_{\alpha+2}(X/m)}\right),$$

where $g_\beta(x) = x^{\beta-1}e^{-x}$ and $G_\beta(x) = \int_0^x t^{\beta-1}e^{-t}dt$. This estimator is easily seen to be the Pitman estimator δ_P of θ under the restriction $\theta \geq m$, i.e. the Bayes estimator of θ with respect to a uniform prior for $\log\theta$ on $[\log m, \infty)$. For the proof of the minimaxity of δ_{vE}, van Eeden uses Lemma 4.2 with a sequence of priors with densities $m^{1/n}/(n\theta^{1+(1/n)})$, $\theta \geq m$, $n = 1, 2, \ldots$. She proves the admissibility of her estimator by using Blyth's (1951) method (see Lemma 4.7).

Note that van Eeden (1995) could have proved the minimaxity of her estimator by using Lemma 4.5 which says that the minimax value of the restricted problem equals the one for the unrestricted problem which equals $1/(1 + \alpha)$. So, all that needs to be shown is that $\sup_{\theta \geq \theta_o} R(\delta_P, \theta) = 1/(1 + \alpha)$ and no sequence of priors needs to be guessed at.

I do not know whether any of Kałuska's other admissible estimators are minimax, nor whether any of his minimax estimators are admissible.

Jafari Jozani, Nematollahi and Shafie (2002) extend the results of van Eeden (1995) to the following exponential-family setting. Let $X = (X_1, \ldots, X_n)'$ have Lebesgue density

$$p_\theta(x) = c(x, n)\tau^{-r\rho}e^{T(x)/\tau^r}\ (x_i > 0, i = 1, \ldots, n), \tag{4.17}$$

where $x = (x_1, \ldots, x_n)$, $\rho > 0$ and $r \neq 0$ are known and $T(x)$ is a sufficient statistic for $\theta = \tau^r$ with a $\Gamma(\rho, \theta)$ distribution. The family (4.17) is a scale parameter family of distributions when $T(ax) = a^r T(x)$ and $c(ax, n) = a^{r\rho - n} c(x, n)$ for all $a > 0$ and all x. The authors show that, for estimating θ under the restriction $\theta \geq m$ with scale-invariant squared-error loss, van Eeden's proofs can be adapted to their case and that

$$\delta(X) = \frac{T(x)}{\rho + 1} \left(1 + \frac{\left(\frac{T(X)}{m} \right)^{\rho+1} e^{-T(X)/m}}{\int_0^{T(X)/m} x^{\rho+1} e^{-x} dx} \right)$$

is an admissible minimax estimator of θ. They note that their results solve this minimax problem for scale parameter estimation in a lognormal, an exponenial, a Rayleigh, a Pareto, a Weibull, a Maxwell and an inverse normal distribution. Here too, minimaxity can be proved by using Lemma 4.5 and showing that the risk function of δ is, for $\theta \in \Theta$, upper-bounded by the minimax value for the unrestricted problem.

The problem of minimax estimation of a lower-bounded scale parameter of an F distribution was considered by van Eeden and Zidek (1994a,b) and by van Eeden (2000). Let X have density

$$f_\theta(x) = \frac{\Gamma(\alpha + n - 1)}{\Gamma(n)\Gamma(\alpha - 1)} \frac{\theta^{\alpha-1} x^{n-1}}{(\theta + x)^{(\alpha+n-1)}} \qquad x > 0,$$

where $n > 0$ and $\alpha > 3$ are given constants and where $\theta \geq m$ for a given $m > 0$. Among many other things, van Eeden and Zidek show that the estimator

$$\delta_{vEZ}(X) = \max \left\{ \frac{\alpha - 3}{n + 1} X, m \right\}$$

is the unique minimax estimator of θ in the class of truncated linear estimators $\mathcal{C} = \{\delta | \delta(X) = \max\{aX, m\}\}$. They also show that the minimax value for this class is given by $(\alpha + n - 2)/((\alpha - 2)(n + 1))$. That this estimator is also minimax among all estimators was shown by van Eeden (2000). She used the results of Gajek and Kałuszka (1995, Section 3) discussed in Section 4.1, to obtain the minimax value for the problem. However, as already mentioned above in the discusssion of these Gajek and Kałuszka results, the minimax value for estimating a lower-bounded scale parameter can be obtained from Lemma 4.5. What is needed is that the Pitman estimator of the unrestricted scale parameter has finite (constant) risk M and that there exists an estimator of $\theta \in R_+^1$ with a risk function $\leq M$ for all $\theta \geq m$. Now, it can easily be seen that this Pitman estimator equals $((\alpha - 3)/(n+1))X$ and that its (constant) risk function equals $(\alpha + n - 2)/((\alpha - 2)(n + 1))$. This method of proof avoids having to check the Gajek–Kałuszka condition on the density f_θ.

This estimator does not solve the minimaxity problem for the lower-bounded scale parameter of an F distribution in a satisfactory manner in the sense that our minimax estimator is inadmissible. This inadmissibility was proved by van Eeden and Zidek (1994a,b) by using results of Charras and van Eeden (1994). I do not know of any result giving an admissible minimax estimator for this case.

Now look at the case where $X_i \sim^{ind} \mathcal{U}(\theta - 1, \theta + 1)$, $i = 1, \ldots, n$, with $\theta \geq 1$ and squared-error loss. Then the unrestricted Pitman estimator δ_o for estimating $\theta \in R^1$ is given by $(Y_1 + Y_2)/2$, where $Y_1 = \min_{1 \leq i \leq n} X_i$ and $Y_2 = \max_{1 \leq i \leq n} X_i$. Its risk function is independent of θ and is given by $2/((n+1)(n+2))$. It is minimax for the unrestricted problem. The Pitman estimator is

$$\delta_P(Y) = \frac{\int_{\max(Y_2-1,1)}^{Y_1+1} t \, dt}{\int_{\max(Y_2-1,1)}^{Y_1+1} dt} = \begin{cases} \frac{1}{2}(Y_1 + Y_2) & \text{when } Y_2 \geq 2 \\ \frac{1}{2}(Y_1 + 2) & \text{when } Y_2 < 2, \end{cases}$$

where $Y = (Y_1, Y_2)$. By Lemma 4.4, the minimax values for the restricted and the unrestricted problems are equal. Further, note that $P_\theta(\delta_o(Y) = \delta_P(Y)) = 1$ for $\theta \geq 3$, so for those θ the estimators have the same risk. When $1 \leq \theta < 3$,

$$R(\delta_o, \theta) - R(\delta_P, \theta) = \frac{(3 - \theta)^{n+2}}{2n(n+1)(n+2)} \geq 0,$$

which shows the minimaxity of the Pitman estimator for the restricted problem.

For the uniform distribution where $X_i \sim^{ind} \mathcal{U}(0, \theta)$, $i = 1, \ldots, n$, $n > 1$, with $\theta \geq 1$ and scale-invariant squared-error loss, the unrestricted Pitman estimator δ_o is $nY/(n-1)$, where $Y = \max_{1 \leq i \leq n} X_i$. It is minimax for the parameter space $\theta > 0$ and its risk function is given by $(n^2 - n + 2)/(n-1)^2(n+1)(n+2)$. The Pitman estimator is easily seen to be given by $n \max(1, Y)/(n-1)$. It is minimax for the restricted problem because

$$R(\delta_o, \theta) - R(\delta_P, \theta) = \frac{n}{\theta^n} \left(\frac{1}{n+2} - \frac{2\theta}{n+1} + \frac{s\theta - 1}{n} \right)$$

$$= \frac{2}{\theta^n} \left(\frac{\theta}{n+1} - \frac{1}{n+2} \right) \geq 0 \Longleftrightarrow \theta \geq \frac{n+1}{n+2}.$$

Two more problems for which admissible minimax estimators are known are the case of a lower-bounded and of an upper-bounded location parameter of an exponential distribution. Let X have density

$$f_\theta(x) = e^{-(x-\theta)} \qquad x > \theta,$$

where $\theta \leq 0$ or $\theta \geq 0$. Berry (1993) obtains admissible minimax estimators of θ for squared error loss for these problems. For the case where $\theta \leq 0$, he uses Lemma 4.2 with the sequence of prior densities $e^{\theta/\nu}/n$, $\theta \leq 0$, $\nu = 1, 2, \ldots$ to show that $\min\{0, x\} - 1$ is minimax. He notes that it is Bayes with respect to the uniform prior on $(-\infty, 0]$, i.e. it is the Pitman estimator. The risk function is given by $1 + 2\theta e^{\theta} \leq 1$ with strict inequality if and only if $\theta < 0$. Berry uses Farrell's (1964) Theorem 7 for its admissibility. When $\theta \geq 0$, Berry uses Farrell's (1964) theorems 6 and 7 with the uniform prior on $(0, \infty)$ to conclude that $x/(1 - e^{-x}) - 1$ is admissible minimax. Note that each of these estimators satisfies (2.3).

These results of Berry (1993) can easily be extended to the case of a sample $X_1, \ldots, X_n$ from their distribution because $Y = \min_{1 \leq i \leq n} X_i$ is sufficient for θ and has density $e^{-(y-\theta)/n}/n$, $y > \theta$. Further, the minimax values for these two problems are known from Lemma 4.4, which says that they are equal to the minimax value for the unrestricted problem. So, here again, all that needs to be shown is that risk functions of the estimators are upper-bounded on Θ by the minimax values for the unrestricted problems.

For this problem of estimating an upper-bounded location parameter of an exponential distribution, based on a sample $X_1, \ldots, X_n$, Parsian and Sanjari Farsipour (1997) use the linex loss function $e^{a(d-\theta)} - a(d - \theta) - 1$ with $a < n$, $a \neq 0$ and $\theta \leq 0$. They show (see Chapter 3, Section 3.6) that, with respect to this loss function, the Pitman estimator δ_P of θ is admissible. They also show that the risk function of the unrestricted Pitman estimator δ_o (still with respect to their loss function) is a constant and that $R(\delta_P, \theta) \leq R(\delta_o, \theta)$ with equality if and only if $\theta = 0$. But this does not prove that δ_P is minimax: Lemma 4.4 does not apply here because the loss function is not squared-error.

The theorems of Farrell (1964) used by Berry (1993) are concerned with minimaxity and admissibility of generalized Bayes estimators of location parameters when $X_1, \ldots, X_n$ are independent with common Lebesgue density $f(x - \theta)$. Farrell's loss functions are strictly convex. In particular he looks at the case where $\theta \geq 0$ and thus gets Katz's (1961) result for a lower-bounded normal mean as a special case. Another example he mentions is the case where, for m an even integer ≥ 4,

$$f(x) = \frac{m - 1}{x^m} I(x \leq -1). \qquad (4.18)$$

He notes that, when $\theta \geq 0$, the Pitman estimator δ_P with respect to squared-error loss has a surprising property. It is, for $n = 1$, given by

$$\delta_P(x) = \begin{cases} \dfrac{-x}{m - 2} & \text{when } x \leq -1 \\[2ex] x + \dfrac{m - 1}{m - 2} & \text{when } x > -1, \end{cases}$$

i.e., according to Farrell's (1964) theorems 7 and 8 this Pitman estimator is admissible and minimax, but it is also non-monotone.

A special case of the density (4.18), namely the case where $m = 4$, is considered by Blumenthal and Cohen (1968b) for the case of two independent samples with ordered location parameters. They have (see Section 4.4) numerical evidence that for squared-error loss the Pitman estimator of the vector is, for that distribution, not minimax.

Remark 4.5. Note that, for each of the above two uniform cases, as well as for the exponential location case and for Farrell's density (4.18), the Pitman estimator satisfies the "extra" restriction (see Chapter 2), i.e., the probability that X is in its estimated support is 1 for all $\theta \in \Theta$.

Finally, suppose that X has a Poisson distribution with mean θ where $\theta \geq m$ for a known $m > 0$ and that we use the loss function $(d - \theta)^2/\theta$. Then the minimax value for the unrestricted problem is well known to be equal to 1 and the minimax estimator for that case is X. This estimator has a constant risk function. In this case neither Lemma 4.4 nor Lemma 4.5 applies because θ is neither a location parameter, nor a scale parameter. Moreover, X does not have a Lebesgue density. In Theorem 4.1 the information inequality (4.7) is used to show that X is $(\mathcal{D}_o, \Theta)$-miminax. The proof of this theorem is analogous to the one given by Lehmann (1983, pp. 267–268) (see also Lehmann and Casella, 1998, p. 327) for the case of a lower-bounded mean of a normal distribution.

Theorem 4.1 *For estimating a Poisson mean θ when $\theta \geq m$ for a known $m > 0$ with loss function $(d - \theta)^2/\theta$ the unrestricted maximum likelihood estimator is $(\mathcal{D}_o, \Theta)$-minimax.*

Proof. Suppose that X is not $(\mathcal{D}_o, \Theta)$-minimax. Then there exists a statistic $\delta(X)$ and an $\varepsilon \in (0, 1)$ such that

$$\mathcal{E}_\theta \frac{(\delta(X) - \theta)^2}{\theta} \leq 1 - \varepsilon \quad \text{for all } \theta \geq m. \tag{4.19}$$

From (4.7) and (4.19) it follows that there exists an $\varepsilon \in (0, 1)$ such that

$$\frac{b^2(\theta)}{\theta} + (b'(\theta) + 1)^2 \leq 1 - \varepsilon \quad \text{for all } \theta \geq m,$$

which implies that, for all $\theta \geq m$,

$$-\sqrt{1 - \varepsilon} - 1 \leq b'(\theta) \leq \sqrt{1 - \varepsilon} - 1;$$

$$-\sqrt{\theta(1 - \varepsilon)} \leq b(\theta) \leq \sqrt{\theta(1 - \varepsilon)}. \tag{4.20}$$

Further, the first inequality in (4.20) gives

$$\left.\begin{aligned} \frac{d}{d\theta}\left(b(\theta) + \sqrt{\theta(1-\varepsilon)}\right) = b'(\theta) + \frac{\sqrt{1-\varepsilon}}{2\sqrt{\theta}} \le \\ \sqrt{1-\varepsilon} - 1 + \frac{\sqrt{1-\varepsilon}}{2\sqrt{\theta}} \quad \text{for all } \theta \ge m. \end{aligned}\right\}$$ (4.21)

Now let θ^* satisfy

$$\theta^* > \max\left(m, \frac{\sqrt{1-\varepsilon}}{1 - \sqrt{1-\varepsilon}}\right),$$

then, by (4.21),

$$\frac{d}{d\theta}\left(b(\theta) + \sqrt{\theta(1-\varepsilon)}\right) <$$

$$-\left(1 - \sqrt{1-\varepsilon}\right) + \frac{1 - \sqrt{1-\varepsilon}}{2} = -\frac{1 - \sqrt{1-\varepsilon}}{2} > 0 \quad \text{for all } \theta > \theta^*.$$

But this contradicts (see the second line of (4.20)) the fact that

$$b(\theta) + \sqrt{\theta(1-\varepsilon)} \ge 0 \quad \text{for all } \theta \ge m.$$

So X is $(\mathcal{D}_o, \Theta)$-minimax and the minimax value for this $(\mathcal{D}_o, \Theta)$ estimation problem equals 1.   ♡

This theorem has the following corollary.

Corollary 4.2 *The minimax value for estimating a lower-bounded mean θ of a Poisson distribution with loss function $(d-\theta)^2/\theta$ equals 1 and the restricted maximum likelihood estimator $\max\{X, \theta_o\}$ is an inadmissible minimax estimator.*

Proof. From Theorem 4.1 it follows that $M(\mathcal{D}_o, \Theta_o) = M(\mathcal{D}_o, \Theta) = 1$. Further, by the completeness of $\mathcal{D}$ in $\mathcal{D}_o$ with respect to Θ, $M(\mathcal{D}_o, \Theta) = M(\mathcal{D}, \Theta)$, proving the first result. That $\max\{X, \theta_o\}$ is minimax then follows from the fact that $\max\{X, \theta_o\}$ dominates X on θ and the inadmissibility of $\max\{X, \theta_o\}$ follows from Brown (1986, Theorem 4.23).   ♡

This leaves us, once again, with only an inadmissible minimax estimator. It seemed to me that the Bayes estimator with respect to the uniform prior on $[m, \infty)$ might be an admissible minimax estimator. The reason behind this idea is that I thought that what works for the lower-bounded normal mean and for the lower-bounded gamma scale parameter, would also work for the lower-bounded Poisson mean. That is, using the truncated version of the prior which gives the unique minimax estimator for the case when θ is not restricted gives a minimax estimator for the restricted case. However, this is not the case as is shown by the following reasoning of Strawderman (1999). First note that the Bayes estimator with respect to the uniform prior on $[m, \infty)$ is given by $\delta(X) = \left(\mathcal{E}\theta^{-1}|X\right)^{-1}$, where the expectation is taken with respect to the posterior distribution of θ. This gives

$$\delta(X) = X + \frac{m^X e^{-X}}{\int_m^\infty t^{X-1} e^{-t} dt} = X + \psi(X),$$

where

$$\psi(x) = \frac{m^x e^{-m}}{\int_m^\infty t^{x-1} e^{-t} dt} = \frac{1}{\int_1^\infty t^{x-1} e^{-m(t-1)} dt}.$$

In order to show that this estimator is not minimax, we need to show that there exists a $\theta \geq m$ for which $\Delta R(\theta) = \mathcal{E}_\theta \left(\delta(X) - \theta\right)^2 - \mathcal{E}_\theta(X - \theta)^2 > 0$. Now note that $\psi(x) \to 0$ as $x \to \infty$. So, using Kubokawa's (1994b) intergral-expression-of-risk method, gives, for t an integer ≥ 0,

$$(X - \theta + \psi(t))^2 - (X - \theta)^2 =$$

$$- \sum_{i=t}^\infty \left[(X - \theta + \psi(i+1))^2 - (X - \theta + \psi(i))^2\right] =$$

$$-2 \sum_{i=t}^\infty (\psi(i+1) - \psi(i)) \left(X - \theta + \frac{\psi(i+1) + \psi(i)}{2}\right).$$

This gives

$$\Delta R(\theta) = -2 \sum_{x=0}^\infty \frac{e^{-\theta} \theta^x}{x!} \sum_{i=x}^\infty (\psi(i+1) - \psi(i)) \left(x - \theta + \frac{\psi(i+1) + \psi(i)}{2}\right)$$

$$= -2 \sum_{i=0}^\infty (\psi(i+1) - \psi(i)) \sum_{x=0}^i \frac{e^{-\theta} \theta^x}{x!} \left(x - \theta + \frac{\psi(i+1) + \psi(i)}{2}\right),$$

where, because $\psi(x)$ is strictly decreasing in x, $\psi(i+1) - \psi(i) < 0$ for all $i = 1, 2, \ldots$ Further, for $\theta = m$,

$$\frac{\psi(i+1) + \psi(i)}{2} > \psi(i+1) = \frac{m^{i+1} e^{-\theta_o}}{\int_m^\infty t^i e^{-t} dt} > \frac{m^{i+1} e^{-m}}{i!} = -\sum_{x=0}^i \frac{m^x e^{-m}}{x!} (x - m),$$

which shows that $\Delta R(\theta_o) > 0$ and thus that $\delta(X)$ is not minimax.

4.4 Minimax results when $k > 1$ and Θ is not bounded

Very few minimaxity results have been obtained for the case where $k > 1$ and Θ is not bounded. Of course, minimaxity results for estimating $\theta = (\theta_1, \ldots, \theta_k)$ when the θ_i are lower- or upper-bounded can be obtained from Lemma 4.3

when minimax estimators of the components of θ based on independent samples are known. Otherwise, the only results I have been able to find are for the case where the loss is squared error and

i)   the θ_i are location parameters, $k = 2$ and $\Theta = \{\theta \mid \theta_1 \leq \theta_2\}$;
ii) $X_i \sim \mathcal{N}(\theta_i, \sigma_i^2)$, $i = 1, \ldots, k$, with the σ_i^2's known but not necessarily equal and $\Theta = \{\theta \mid \theta_1 \leq \ldots \leq \theta_k\}$;
iii) $X_i \sim^{ind} \mathcal{N}(\theta_i, 1)$, $i = 1, \ldots, k$ and Θ is a closed, convex subset of R^k with an apex;
iv) $X_i \sim^{ind} \mathcal{N}(\theta_i, \sigma_i^2)$, $i = 1, 2$, $\Theta = \{\theta \mid |\theta_2 - \theta_1| \leq c\}$ for a given positive constant c and known σ_i^2.

Results concerning point i) have been obtained by Blumenthal and Cohen (1968b) for the case where X_i, $i = 1, 2$ are independent and X_i has Lebesgue density $f(x - \theta_i)$, $i = 1, 2$. This is a case where the conditions of Θ of Lemma 4.4 are satisfied, so the restricted and unrestricted mimimax values are equal. They give sufficient conditions on f for the Pitman estimator to be minimax. The normal, uniform and exponential densities satisfy these conditions. Their sufficient conditions for the Pitman estimator to be admissible are satisfied by the normal and exponential densities. And for the density $f(x) = (3/x^4)I(x \leq -1)$ they have numerical evidence that the Pitman estimator is not minimax. The Blumenthal–Cohen (1968b) proofs, which are rather complicated, are based on results of Blumenthal and Cohen (1968a), Farrell (1964) and James and Stein (1961).

Remark 4.6. The density $f(x) = (3/x^4)I(x \leq -1)$ is a member of the family (4.18) for which Farrell (1964) finds (see Section 4.3) that, for squared-error loss, the Pitman estimator of a lower-bounded location parameter is non-monotone, admissible and minimax.

Results concerning point ii) above have been obtained by Kumar and Sharma (1988, 1989, 1993). Most of their results hold for the case where $X_i \sim \mathcal{N}(\theta_i, 1)$, $i = 1, \ldots, k$, $k \geq 2$, squared-error loss and $\theta_1 \leq \ldots \leq \theta_k$. This is again a case where the conditions on Θ of Lemma 4.4 are satisfied and they show the Pitman estimator to be minimax. Further, for $k = 2$, they define so-called mixed estimators, which are given by

$$
\delta(X_1, X_2) = \begin{cases} \begin{pmatrix} X_1 \\ X_2 \end{pmatrix} & \text{when } X_1 \leq X_2 \\[2ex] \begin{pmatrix} \alpha X_1 + (1 - \alpha)X_2 \\ (1 - \alpha)X_1 + \alpha X_2 \end{pmatrix} & \text{when } X_1 > X_2, \end{cases}
$$

where $0 \leq \alpha \leq 1$. These estimators are in $\mathcal{D}$ if and only if $\alpha \leq 1/2$. Kumar and Sharma, in their 1988 paper, show those for $\alpha \leq 1/2$ to be admissible among themselves. They claim to show that the mixed estimators are inadmissible by showing that they are not generalized Bayes, but in fact they only show that they are not generalized Bayes with respect to a prior which has a Lebesgue density. Kumar and Sharma also consider the case where the X_i are normal but do not necessarily have the same variance, as well as the more general case where the densities of the X_i do not necessarily have the same shape. In their (1993) paper they state (see their p. 233) that the Pitman estimator is inadmissible. But Blumenthal and Cohen (1968b) show it to be admissible when $k = 2$. Maybe Kumar and Sharma (1988) forgot to say $k \geq 3$.

For the normal-mean case with $k = 2$ and $\theta_1 \leq \theta_2$, Katz (1963) gives proofs of the admissibility and minimaxity of the Pitman estimator of (θ_1, θ_2). But, as both Blumenthal and Cohen (1968b) and Kumar and Sharma (1988) note, these proofs are "inadequate" – a statement with which I agree.

Concerning point iii) above, Hartigan (2004) shows that, for $X \sim \mathcal{N}_k(\theta, I)$ and squared-error loss, the Pitman estimator δ_P satisfies $R(\delta_P, \theta) \leq k$ for all $\theta \in \Theta$ with equality if and only if θ is an apex of Θ. Then, if the conditions on Θ of Lemma 4.4 are satisfied, δ_P is minimax. And the parameter spaces of Blumenthal–Cohen and Kumar–Sharma do have an apex, so their minimaxity results for $X_i \sim^{ind} \mathcal{N}(\theta_i, 1)$, $i = 1, \ldots, k$ follow from this Hartigan result.

Concerning point iv) above, a new result on minimax estimation of restricted normal means is contained in the following theorem. It is a case where the conditions of Lemma 4.4 are not satisfied.

Theorem 4.2 Let X_1 and X_2 be independent random variables with, for $i = 1, 2$, $X_i \sim \mathcal{N}(\theta_i, \sigma_i^2)$ where the σ_i are known and $|\theta_2 - \theta_1| \leq c$ for a known $c > 0$ satisfying $c \leq m_o \sqrt{\sigma_1^2 + \sigma_2^2}$, where $m_o \approx 1.056742$ is the Casella–Strawderman constant. Then, for squared-error loss, a minimax estimator $\delta(X_1, X_2) = (\delta_1(X_1, X_2), \delta_2(X_1, X_2))$ of the vector $\theta = (\theta_1, \theta_2)$ is given by

$$\delta_1(X_1, X_2) = \frac{1}{1 + \tau}\left(\tau X_1 + X_2 - c \tanh\left(\frac{c}{\sigma^2}(X_2 - X_1)\right)\right)$$

$$\delta_2(X_1, X_2) = \frac{1}{1 + \tau}\left(\tau X_1 + X_2 + \tau c \tanh\left(\frac{c}{\sigma^2}(X_2 - X_1)\right)\right),$$

where $\tau = \sigma_2^2/\sigma_1^2$ and $\sigma^2 = \sigma_1^2 + \sigma_2^2$.

The minimax value for the problem is given by

$$\frac{2\sigma_1^2 \sigma_2^2}{\sigma^2} + \left(1 + \frac{\sigma_2^4}{\sigma_1^4}\right)\frac{\sigma_1^4}{\sigma^2} \sup_{|\Delta| \leq m} \mathcal{E}\left(m \tanh mZ - \Delta\right)^2,$$

where $Z \sim \mathcal{N}(\Delta, 1)$ and $m = c/\sigma$. Note that, from Hartigan (2004), we know that when $\sigma_1^2 = \sigma_2^2 = \lambda^2$, this minimax value is $< 2\lambda^2$.

Proof. The proof presented here uses a technique used by van Eeden and Zidek (2004). They study the problem of estimating θ_1 based on (X_1, X_2) when $|\theta_2 - \theta_1| \leq c$. As will be seen in the next chapter, they obtain for this model an estimator of θ_1 based on (X_1, X_2), which is admissible and minimax. In their proof they use the following rotation technique of Blumenthal and Cohen (1968a) (see also Cohen and Sackrowitz, 1970). Let

$$Y_1 = \frac{\tau X_1 + X_2}{1 + \tau} \qquad Y_2 = \frac{-X_1 + X_2}{1 + \tau}$$

$$\mu_1 = \mathcal{E}_\theta Y_1 = \frac{\tau \theta_1 + \theta_2}{1 + \tau} \qquad \mu_2 = \mathcal{E}_\theta Y_2 = \frac{-\theta_1 + \theta_2}{1 + \tau}.$$

(4.22)

Then $|\theta_2 - \theta_1| \leq c$ if and only if $|\mu_2| \leq c/(1+\tau)$ and μ_1 is unrestricted. Further note that

$$X_1 = Y_1 - Y_2 \qquad X_2 = Y_1 + \tau Y_2$$

(4.23)

$$\theta_1 = \mu_1 - \mu_2 \qquad \theta_2 = \mu_1 + \tau \mu_2$$

and that Y_1 and Y_2 are independent normal random variables. So, finding a minimax estimator of θ based on (X_1, X_2) under the restriction $|\theta_2 - \theta_1| \leq c$ is equivalent to finding a minimax estimator of $(\mu_1 - \mu_2, \mu_1 + \tau \mu_2)$ based on (Y_1, Y_2) under the restriction $|\mu_2| \leq c/(1 + \tau)$. To solve this problem, take a sequence λ_n, $n = 1, 2, \ldots$, of priors for $\mu = (\mu_1, \mu_2)$ where, for each n, μ_1 and μ_2 are independent, with $\mu_1 \sim \mathcal{N}(0, n)$ and the prior for μ_2 with mass $1/2$ on each of $\pm c/(1 + \tau)$. The independence of μ_1 and μ_2 combined with the conditional, given μ, independence of Y_1 and Y_2 implies that, for $i = 1, 2$, the Bayes estimator $\delta_{n,i}(Y_1, Y_2)$ of μ_i depends on Y_i only and

$$\delta_{n,1}(Y_1, Y_2) = \delta_{n,1}(Y_1) = \frac{Y_1}{1 + (\gamma^2/n)}$$

$$\delta_{n,2}(Y_1, Y_2) = \delta_2(Y_2) = \frac{c}{1 + \tau} \tanh\left(\frac{c}{\sigma_1^2} Y_2\right),$$

where $\gamma^2 = Var(Y_1) = \sigma_1^2 \sigma_2^2 / \sigma^2$. The components $\delta_{n,B,i}(Y_1, Y_2)$, $i = 1, 2$, of the Bayes estimator $\delta_{n,B}(Y_1, Y_2)$ of $(\mu_1 - \mu_2, \mu_1 + \tau \mu_2)$ are then given by (see 4.23)

$$\delta_{n,B,1}(Y_1, Y_2) = \frac{Y_1}{1 + (\gamma^2/n)} - \frac{c}{1 + \tau} \tanh\left(\frac{c}{\sigma_1^2} Y_2\right)$$

and

$$\delta_{n,B,2}(Y_1, Y_2) = \frac{Y_1}{1 + (\gamma^2/n)} + \frac{\tau}{1 + \tau} c \tanh\left(\frac{c}{\sigma_1^2} Y_2\right).$$

Because of the independence of the Y_i, the MSE of $\delta_{n,B}$ as an estimator of $(\mu_1 - \mu_2, \mu_1 + \tau \mu_2)$ is given by is given by

$$2\left(\frac{2\sigma_1^2 \sigma_2^2 / \sigma^2}{(1 + (\gamma^2/n))^2} + \mu_1^2 \frac{\gamma^4}{(\gamma^2 + n)^2}\right) + (1 + \tau^2)\mathcal{E}_{\mu_2}(\delta_2(Y_2) - \mu_2)^2. \qquad (4.24)$$

By Casella and Strawderman (1981) we know that, when $c \leq m_o \sigma$, the Bayes risk of $\delta_2(Y_2)$ as an estimator of μ_2 based on Y_2 under the restriction $|\mu_2| \leq c/(1+\tau)$ equals $\sup(\mathcal{E}_{\mu_2}(\delta_2(Y_2) - \mu_2)^2| \ |\mu_2| \leq c/(1+\tau))$. Further, given that the prior for μ_1 is $\mathcal{N}(0, n)$, the Bayes risk of the second factor in the first term of (4.24) converges, as $n \to \infty$, to $\sigma_1^2 \sigma_2^2/\sigma^2$, which is the (constant) risk function of Y_1. The minimaxity then follows from Lemma 4.2 and (4.22).

From (4.24) the minimax value for the problem is found to be

$$2\frac{\sigma_1^2 \sigma_2^2}{\sigma^2} + (1 + \tau^2)r_2,$$

where

$$r_2 = \sup \left(\mathcal{E}_{\mu_2} \left(\frac{c}{1+\tau} \tanh \left(\frac{c}{\sigma_1^2} Y_2 \right) - \mu_2 \right)^2 | \ |\mu_2| \leq c/(1+\tau) \right)$$

$$= \sup_{|\theta_2 - \theta_1| \leq c} \mathcal{E}_\theta \left(\frac{c}{1+\tau} \tanh \frac{c}{\sigma^2}(X_2 - X_1) - \frac{\theta_2 - \theta_1}{1+\tau} \right)^2$$

$$= \frac{\sigma_1^4}{\sigma^2} \sup_{|\Delta| \leq c/\sigma} \mathcal{E} \left(\frac{c}{\sigma} \tanh \left(\frac{c}{\sigma} Z \right) - \Delta \right)^2,$$

where $Z \sim \mathcal{N}(\Delta, 1)$. This proves the result concerning the minimax value. $\heartsuit$

Remark 4.7. For $X \sim \mathcal{N}_k(\theta, \Sigma)$ with $\theta_i \geq 0$, $i = 1, \ldots, k$, Sengupta and Sen (1991, Theorem 4.2) seem to say that the MLE of θ is minimax. I do agree with this when Σ is diagonal (see Lemma 4.3). However, I do not see why this is true when Σ is not diagonal.

4.5 Discussion and open problems

It is clear from the above that (admissible) minimax estimators have been found only for relatively few cases of restricted parameter spaces and mostly only for (scale-invariant) squared-error loss. When Θ is bounded, most results hold only for "small" parameter spaces. When Θ is not bounded, the known minimax estimators are often inadmissible.

Given how difficult it is to find minimax estimators, one wonders whether they are "worth the trouble": i.e., can one with less effort find estimators which are "almost as good"? Some, mostly numerical, results have been obtained to help answer this question. For the bounded-normal-mean problem with $X \sim \mathcal{N}(\theta, 1)$ and $-m \leq \theta \leq m$, Gatsonis, MacGibbon and Strawderman (1987) compare, among other things, the Casella–Strawderman–Zinzius minimax estimator of θ with the (admissible) Pitman estimator δ_P. For $m = .5$, e.g., the minimax value for the problem is $\sim .199$ (see Casella and Strawderman, 1981; or Brown and Low, 1991), while from their graphs one sees that the maximum difference between the two risk functions is $\sim .05$. Similar results hold for the other two values of m they have results for, namely, $m = .75$ and $m = 1.5$. Further, in each of these three cases, numerical results show that δ_P dominates the minimax estimator over a large part of the interval $[-m, m]$, while the authors show that the MLE dominates the minimax estimator over an interval of the form $[-m^*, m^*]$ with numerical evidence that $m^* \approx .75\ m$. They moreover show that the risk function of the Pitman estimator is ≤ 1, the minimax value $M(\mathcal{D}_o, \Theta_o)$ for the unrestricted problem, a result that is generalized by Hartigan (2004) to $k \geq 1$ with Θ a closed, convex subset of R^k with a non-empty interior. He shows (as already mentioned in Section 4.1) that for that case equality holds if and only if θ is an apex of Θ. Note that all the above risk-function values from the Gatsonis, MacGibbon and Strawderman (1987) paper have been read off their graphs and are thus very approximate.

A class of "nearly minimax" estimators considered for the bounded-normal-mean problem with $X \sim \mathcal{N}(\theta, 1)$, $-m \leq \theta \leq m$ and squared-error loss, is the class of linear minimax estimators, i.e., estimators which are minimax among linear estimators. Results for this problem have been obtained by Donoho, Liu and MacGibbon (1990). They show that the linear minimax estimator is given by $\delta_L(X) = (m^2/(m^2 + 1))X$ and that the minimax linear risk is given by $\rho_L(m) = m^2/(m^2 + 1)$. They study the properties of $\lambda(m) = \rho_L(m)/\rho(m)$, where $\rho(m)$ is the minimax value for the problem. One of their (many) results says that $\lambda(m) \leq 1.25$ for all m and they quote numerical results of Feldman from his unpublished thesis stating that $1.246 \leq \lambda(m) \leq 1.247$. Further, Gourdin, Jaumard and MacGibbon (1990) show that $\lambda(m) \in [1.246408, 1,246805]$. So, risk-function-wise this estimator is not a bad alternative to the minimax estimator. Donoho, Liu and MacGibbon (1990) also study the estimator $\delta^*(X) = XI(m \geq 1)$, which has $\sup_{-m \leq \theta \leq m} R(\delta^*, \theta) = \min(m^2, 1)$ and $\max_{m>0}(\min(m^2, 1))/\rho(m) \approx 2.22$. So, risk-function-wise, δ_L is to be preferred over δ^*. The authors also have many results for the case where $k \geq 2$ for various Θ.

Another class of "nearly minimax" estimators for the above bounded-normal-mean problem can be found in Vidakovic and DasGupta (1996). They consider so-called Γ-minimax linear estimators. In the Γ-minimax approach one specifies a class Γ of priors (in this case the authors took all symmetric unimodal ones on Θ). Then a Γ-minimax estimator minimizes $\sup_{\gamma \in \Gamma} r_\gamma(\delta)$, where $r_\gamma(\delta)$ is the Bayes risk of the estimator δ for the prior γ – all of this of course for a given loss function which in the present case is squared-error. For more on this Γ-minimax approach, see, e.g., Berger (1984). Vidakovic and DasGupta (1996) use this approach but restrict themselves to linear estimators and show that $\delta_L^*(X) = (m^2/(m^2 + 3))X$ with corresponding Γ-minimax risk $m^2/(m^2 + 3)$. This Γ-minimax risk satisfies $\rho_L(m)/\rho(m) \leq 1.074$, showing that δ^* is better that δ_L in the minimax sense. The difference is surprising, given how close the estimators are for large m. These authors also have results for $k \geq 2$.

Linear minimax estimators have also been studied for the case where $X \sim$ Poisson with mean θ, $\theta \in [0, m]$ and loss function $L(d, \theta) = (d - \theta)^2/\theta$. Johnstone and MacGibbon (1992) show that this estimator is given by $\delta^{**}(X) = (m/(m+1))X$ with linear minimax value $m/(m+1)$. From numerical results they find that $\lambda(m) \leq 1.251$, while Gourdin, Jaumard and MacGibbon (1990) show that $\lambda \in [1.250726, 1.250926]$. Johnstone and MacGibbon (1992) comment on the surprising similarity of these bounds with those for the bounded-normal-mean case.

Marchand and MacGibbon (2000) make comparisons between various estimators for the binomial case with $\theta \in [m, 1 - m]$ for $0 < m < 1/2$, as well as with $\theta \in [0, b]$ with $0 < b < 1$. For estimating θ they have results for squared-error loss and for the loss function $(\theta - d)^2/(\theta(1 - \theta))$. One of the estimators they consider is the linear minimax estimator. Their numerical results are presented in the form of graphs of the risk function.

Remark 4.8. The above linear minimax estimators are not estimators in the sense of this monograph – they do not satisfy (2.3). The only comment I have been able to find about this is in Vidakovic and DasGupta (1996). They give, for the bouded-normal-mean problem, numerical values of the infimum over Θ of the probability that their estimator is in Θ. For $k = 1$, e.g., they find $\approx .84$ for $m = 3$ and $\approx .52$ for $m = 50$. They also derive estimators of the form $\delta(X) = cX$ for which this infimum is $\geq 1 - \alpha$ for a given α. But none of the above proposers of linear minimax estimators seem to look at the properties of dominators of their linear estimators which satisfy (2.3).

On the question of how much can be gained minimax-wise: as has been seen already, nothing can be gained when $k = 1$ and θ is a lower-bounded location or scale parameter and, more generally, when $k > 1$ and Θ satisfies the conditions of Lemma 4.4 or of Lemma 4.5. Another case where the two minimax values are equal is a symmetrically truncated binomial parameter when the difference between Θ_o and Θ is small. Note that these two situations are very different. In the case of location and scale parameters, the two minimax estimators δ_o and δ of θ for, respectively, the parameter spaces Θ_o and Θ are different but have $\sup_{\theta \in \Theta_o} R(\delta_o, \theta) = \sup_{\theta \in \Theta} R(\delta, \theta)$. In the binomial case, the minimax estimator for $(\mathcal{D}_o, \Theta_o)$ satisfies (2.3) and has a constant risk function so that it is also $(\mathcal{D}, \Theta)$-minimax.

On the question of how different $M(\mathcal{D}_o, \Theta_o)$ and $M(\mathcal{D}, \Theta)$ are when they are not equal, numerical results have been obtained for the bounded-normal-mean problem as well as for a symmetrically restricted binomial probability.

For $X \sim \mathcal{N}_k(\theta, I)$ with squared-error loss and $\Theta = \{\theta \mid \sum_{i=1}^{k} \theta_i^2 \leq m^2\}$, minimax values have been obtained by Casella and Strawderman (1981) and by Brown and Low (1991) for $k = 1$. Their results are summarized in Tab 4.1.

Table 4.1. Minimax values, $X \sim \mathcal{N}(\theta, 1)$, $|\theta| \leq m$.

m	.100	.200	.300	.400	.500	.600	.700
$M(\mathcal{D}, \Theta)$	.010	.038	.083	.138	.199	.262	.321

m	.800	.900	1.00	2.00	3.00	.500	10.0
$M(\mathcal{D}, \Theta)$	.374	.417	.450	.645	.751	.857	.945

Results from Berry (1990) for this normal-mean problem for $k = 2$ and for $k = 3$ are given in Tab 4.2 and Tab 4.3, respectively. The starred values in the Berry tables are (approximately) the largest m for which the Bayes estimator with respect to the uniform prior on the boundary of Θ is minimax and the value .260 for $k = 3$ and $m = .800$ might well be incorrect.

Table 4.2. Minimax values, $X \sim \mathcal{N}_k(\theta, 1)$, $\sum_{i=1}^{k} \theta_i^2 \leq m^2$, $k = 2$.

m	.200	.400	.600	.800	1.00	1.20	1.40	1.53499*
$M(\mathcal{D}, \Theta)$	.039	.148	.305	.482	.655	.806	.927	.989

Table 4.3. Minimax values, $X \sim \mathcal{N}_k(\theta, I)$, $\sum_{i=1}^{k} \theta_i^2 \leq m^2$, $k = 3$.

m	.200	.400	.600	.800	1.00	1.20	1.40	1.60	1.80	1.90799*
$M(\mathcal{D}, \Theta)$	.039	.152	.321	.260	.746	.961	1.158	1.330	1.473	1.538

All values in these three tables are of course $< k$ and, for fixed k, increasing in m. Further, the relative gain in minimax value from restricting the parameter space increases in k for fixed m. Clearly, substantial gains can be made.

For $X \sim N_k(\theta, I)$ with $\Theta = \{\theta \mid \theta_i \in [-m_i, m_i], i = 1, \ldots, k\}$, the minimax value equals the sum of the minimax values for the component problems. This implies, e.g., that the relative gain in minimax value for such Θ is independent of k when m_i is independent of i.

Brown and Low (1991) also give lower bounds for $M(\mathcal{D}, \Theta)$. One of these bounds says that $M(\mathcal{D}, \Theta) \geq (1 + (\pi^2/m^2))^{-1}$, a bound which was earlier obtained by Klaassen (1989) from his spread-inequality.

For the binomial case many minimax values have been obtained, for squared-error loss as well as for the normalized loss function $(d - \theta)^2/(\theta(1 - \theta))$. Tab 4.4 contains some of those values for squared-error loss, $\Theta_o = [0, 1]$ and $m \leq \theta \leq 1 - m$. They are taken from Moors (1985).

Table 4.4. Minimax values, $X \sim \text{Bin}(n, \theta)$, $m \leq \theta \leq 1 - m$.

m	.35	.15	.05	.00
$n = 3$	.0173	.0329	.0335	.0335
$n = 10$	.0104	.0143	.0144	.0144
$n = 15$	.0078	.0104	.0105	.0105

For each of the values of n in Tab 4.4 we have $(n + \sqrt{n}/2)/(n + \sqrt{n}) \leq 1 - m$ for $m = .05$, implying, as already mentioned above, that the minimax value for this value of m equals the minimax value for $m = 0$, namely, $(4(1 + \sqrt{n})^2)^{-1}$. That is, for this case the minimax value reaches its maximum over m for a Θ, which is smaller than Θ_o. For n and m such that $(n + \sqrt{n}/2)/(n + \sqrt{n}) > 1 - m$, the minimax value is strictly decreasing in m and in n and the gain in minimax value as a percentage of $M(\mathcal{D}_o, \Theta_o)$ increases in m and decreases in n – all this in accord with intuition.

5

Presence of nuisance parameters

In this chapter results are presented on (in)admissibility and minimaxity when nuisance parameters are present. In almost all of the published results on this problem the following models are considered. Let $X_{i,j}, j = 1, \ldots, n_i, i = 1, \ldots, k$, be independent random variables where, for $i = 1, \ldots, k$, the $X_{i,j}$ are identically distributed with distribution function $F_i(x; \mu_i, \nu_i)$. Then the estimation problem is one of the following:

i) $\nu = (\nu_1, \ldots, \nu_k)$ is known, $\mu = (\mu_1, \ldots, \mu_k)$ is unknown, restrictions are imposed on μ and, for a given $i_o \in \{1, \ldots, k\}$, $\theta_1 = \mu_{i_o}$ is to be estimated with $\lambda = (\mu_i, i \neq i_o)$ as a vector of nuisance parameters. In the notation of Chapter 2, $M = 1$ and $K = k$;

ii) μ and ν are both unknown, restrictions are imposed on μ and, for a given $i_o \in \{1, \ldots, k\}$, $\theta_1 = \mu_{i_o}$ is to be estimated with $\lambda = (\nu, \mu_i, i \neq i_o)$ as a vector of nuisance parameters. In the notation of Chapter 2, $M = 1$ and $K = 2k$;

iii) μ and ν are both unknown, restrictions are imposed on (μ, ν) and, for a given $i_o \in \{1, \ldots, k\}$, (μ_{i_o}, ν_{i_o}) is to be estimated with $\lambda = ((\mu_i, \nu_i), i \neq i_o)$ as a vector of nuisance parameters. In the notation of Chapter 2, $M = 2$ and $K = 2k$.

In some cases where the above models are studied, a linear combination of the parameters of interest is the estimand. Various other variations on the above models are also studied.

In all cases, the resulting parameter space is denoted by Ω, Θ is defined as in (2.1), estimators are based on $X = \{X_{i,j}, j = 1, \ldots, n_i, i = 1, \ldots, k\}$ and satisfy (2.3).

For the models described in i) - iii), many results have been obtained on comparisons between estimators based on $X^* = \{X_{i,j} \mid j = 1, \ldots, n_i, i = i_o\}$ and those based on X. As will be seen, it is often possible to find estimators based on X which dominate a "best" one based on X^*. For instance, when

$X_i \sim^{ind} \mathcal{N}(\mu_i, 1)$, $i = 1, 2$, with $\mu_1 \leq \mu_2$ and squared-error loss is used to estimate $\theta_1 = \mu_1$ with μ_2 as a nuisance parameter, the MLE of μ_1 based on $X = (X_1, X_2)$ (i.e., the first component of the MLE of (μ_1, μ_2) under the restriction $\mu_1 \leq \mu_2$) dominates $X^* = X_1$ on $\Omega = \{(\mu_1, \mu_2) \mid \mu_1 \leq \mu_2\}$. So, using both X_1 and X_2 to estimate $\mu_1 \in \Theta = R^1$ leads to an improved estimator of the parameter of interest. As a second example, suppose that the $X_{i,j}$ are $\mathcal{N}(\mu, \nu_i^2)$ with μ the parameter of interest and the ν_i^2 unknown and satisfying $\nu_1^2 \leq \ldots \leq \nu_k^2$. Then, as will be seen later, the so-called Graybill-Deal estimator of $\mu \in \Theta = R^1$ is universally inadmissible (with respect to the class of loss functions which are nondecreasing in $|d - \mu|$) on $\Omega = \{\mu, \nu_1^2, \ldots, \nu_k^2 \mid -\infty < \mu < \infty, \nu_1^2 \leq \ldots \leq \nu_k^2\}$. This is an example of a case where putting restrictions only on the nuisance parameters makes it possible to improve on the estimation of the parameter of interest. In each of these two examples Θ is the real line.

Another question that is considered for this kind of problem is whether the improved estimators are themselves admissible and, if not, are (admissible, minimax) dominators available?

For solving the above-described kinds of problems, some authors use the techniques of Brewster and Zidek (1974). These techniques can be described as follows. Under very general conditions on the family of distributions $\mathcal{F} = \{F_\gamma \mid \gamma \in \Gamma\}$ of an observable random vector X and for a strictly bowl-shaped loss function, Brewster and Zidek give three ways of obtaining dominators of an equivariant estimator δ of (a subvector of) γ. In their first method they condition on an appropriately chosen statistic $T = T(X)$ and obtain dominators by studying $\mathcal{E}_\gamma[\mathcal{E}_\gamma(L(\delta, \gamma) \mid T)]$ as a function of δ and γ for $\gamma \in \Gamma$. Their second method consists of taking the limit of an appropriately chosen sequence of testimators, while their third method is a modification of their second method as described on page 34 (lines -8 to -5) of their paper. Brewster and Zidek (1974) give several examples of their techniques, among which are two where Γ is a restricted parameter space. Both examples are concerned with simply-tree-ordered parameters. In the first of these examples these parameters are normal means; in the second one they are normal variances. Later in this chapter, these Brewster–Zidek dominators are compared with results of other authors for these two problems and, as will be seen, some authors obtain the Brewster–Zidek dominators by a different method, but do not refer to the Brewster–Zidek results.

For several of the above-described problems, authors only consider the $(\mathcal{D}_o, \Omega)$-case, implying that some their "estimators" satisfy (2.5), but not (2.3), while they compare their "estimators" on Ω. Solutions to such problems are presented in this chapter because, as already noted earlier, such results are often useful for solving related $(\mathcal{D}, \Omega)$-problems.

Results for the case where $\nu = (\nu_1, \ldots, \nu_k)$ is known are given in Section 5.1 for location problems and in Section 5.2 for scale problems. Those for the case where ν is unknown can be found in Section 5.3. In each of these three sections Ω is, with a few exceptions, defined by inequalities among the parameters. Section 5.4 contains results for restrictions in the form of more general cones, in particular polygonal cones, orthant cones and circular cones. Some (admissible) minimax estimators are given in Section 5.5.

5.1 Location parameter estimation with known ν

In this section we suppose that $\nu = (\nu_1, \ldots, \nu_k)$ is known and location parameters are to be estimated.

We first consider the case where $X_i \sim^{ind} \mathcal{N}(\mu_i, \nu_i^2)$ with known ν_i^2's, $\Omega = \{\mu \mid \mu_1 \leq \ldots \leq \mu_k\}$ and squared-error loss. Let, for some given $i \in \{1, \ldots, k\}$, $\theta_1 = \mu_i$ be the parameter of interest. Then $\Theta = R^1$ and for squared-error loss the best estimator based on X_i alone is of course X_i. But, as Lee (1981) shows, the MLE $\hat{\mu}_i$ of μ_i (i.e., the i-th component of the MLE $\hat{\mu} = (\hat{\mu}_1, \ldots, \hat{\mu}_k)$ of μ) dominates X_i. A stronger result was obtained by Kelly (1989). He shows that, with respect to the class of loss functions which are non-constant and non-decreasing in $|d - \mu_i|$, $\hat{\mu}_i$ universally dominates X_i and this result was, for $k = 2$, proved by Kushary and Cohen (1989) for more general location families. However (see Garren, 2000), the Kelly result does not hold when the ν_i^2's are unknown and, in the MLE of μ_i, $\nu_1^2, \ldots, \nu_k^2$ are replaced by their unrestricted MLEs. In fact, Garren shows that X_i and this "plug-in" estimator are non-comparable for squared-error loss. But Hwang and Peddada (1994) show that when the $\nu_i^2 = \nu^2$, $i = 1, \ldots, k$ with ν^2 unknown, then Kelly's universal domination result still holds when, in the MLE of μ_i, ν^2 is replaced by its usual pooled estimator. Also, Lee's result does not imply that $c'\hat{\mu}$ dominates $c'X$ as an estimator of $c'\mu$ for vectors $c \neq (0, \ldots, 0, 1, 0, \ldots, 0)$. In fact, Fernández, Rueda and Salvador (1999) show that, when c is the so-called central direction of the cone Ω, then, for large enough k, $c'X$ has a smaller mean-squared error than $c'\hat{\mu}$ when $\mu_1 = \ldots = \mu_k = 0$. This central direction of a cone (see Abelson and Tukey, 1963) is the direction which minimizes the maximum angle with the directions in the cone. Further, Gupta and Singh (1992) show that, when $k = 2$ and $\nu_1 = \nu_2$, $\hat{\mu}_i$ dominates X_i, for $i = 1, 2$, also by the Pitman-closeness criterion.

Remark 5.1. For $k = 2$, Lee's (1981) result, as well as Kelly's (1989) result for loss functions which are strictly increasing in $|d - \mu_i|$, are special cases of Brewster and Zidek's (1974) Theorem 2.2.1. Neither Lee nor Kelly seems to have been aware of this Brewster–Zidek result.

The above results of Lee and Kelly for simply ordered normal means do not necessarily hold for incomplete orderings like, e.g., the simple-tree ordering given by $\mu_1 \leq \mu_i$, $i = 1, \ldots, k$. For this ordering Lee (1988) considers the case where $X_i \sim^{ind} \mathcal{N}(\mu_i, \nu_i^2)$, $i = 1, \ldots, k$ with squared-error loss and the ν_i^2 known. He compares X_i with the MLE $\hat{\mu}_i$ of μ_i and shows that, for $i \geq 2$, $\hat{\mu}_i$ dominates X_i on $\Omega = \{\mu \mid \mu_1 \leq \mu_i, i = 2, \ldots, k\}$ when $\nu_1 \leq \nu_i$, $i = 2, \ldots, k$. However, for estimating μ_1, Lee shows that, when μ_i and ν_i are, respectively, upper- and lower-bounded as $k \to \infty$, X_1 has a smaller MSE than $\hat{\mu}_1$ for k large enough, whereas for $\mu_1, \ldots, \mu_k$ and $\nu_2, \ldots, \nu_k$ fixed, $\hat{\mu}_1$ has a smaller MSE than X_1 for small enough ν_1^2. A related result for this normal-mean problem can be found in Fernández, Rueda and Salvador (1999). For the simple-tree order, e.g., they show that when $\nu_1 = \ldots = \nu_k$ and the μ_i are bounded as $k \to \infty$, $c'X$ has a smaller MSE than $c'\hat{\mu}$ for sufficiently large k when c is the central direction of the cone Ω, i.e. $c = (-(k-1), 1, \ldots, 1)$ (see, e.g., Robertson, Wright and Dykstra, 1988, p. 181).

Three other examples (none of them a location problem) where a component of the restricted MLE does not dominate the corresponding component of the unrestricted one, are Poisson, uniform and binomial cases with a simple ordering of the parameters and $k = 2$.

Kushary and Cohen (1991) obtain results for $X_i \sim^{ind} \text{Poisson}(\mu_i)$, $i = 1, 2$, with $0 < \mu_1 \leq \mu_2$ and squared-error loss. They show that for estimating μ_1, X_1 is dominated by the MLE of μ_1, whereas for estimating μ_2, if $\delta(X_2)$ is admissible among estimators based on X_2 alone, it is admissible among estimators based on (X_1, X_2). Parsian and Nematollahi (1995) show that this Kushary–Cohen result concerning the estimation of μ_2 holds for the more general case of a strictly convex loss function. For the estimation of μ_1 Parsian and Nematollahi show that, for the entropy loss function $L(d, \mu)$, which for estimating a Poison mean μ, satisfies $\mu L(d, \mu) = d/\mu - \log(d/\mu) - 1$, $X_1 + 1$ (which is admissible for estimating μ_1 when X_2 is not observed) is inadmissible when X_2 is observed.

For the case where $X_{i,j} \sim^{ind} \mathcal{U}(0, \mu_i)$, $j = 1, \ldots, n_i$, $i = 1, 2$, $\mu_1 \leq \mu_2$, the restricted and unrestricted MLEs of μ_1 are equal, while, for estimating μ_2, the restricted MLE dominates the unrestricted one (see Section 5.2).

For the binomial case with $X_i \sim^{ind} \text{Bin}(n_i, \mu_i)$, $i = 1, 2$ and $\mu_1 \leq \mu_2$, Hengartner (1999) shows that, for estimating μ_2, X_2/n_2 and the MLE $\hat{\mu}_2$ are noncomparable for squared-error loss. For $n_1 = 1$, e.g., he shows that

$$\mathcal{E}_\mu \left(\frac{X_2}{n_2} - \mu_2 \right)^2 > \mathcal{E}_\mu \left(\hat{\mu}_2 - \mu_2 \right)^2 \iff \frac{n_2}{3n_2 + 1} < \mu_2 < 1.$$

Remark 5.2. Sampson, Singh and Whitaker (2003) say (their p. 300) that Kushary and Cohen (1991) establish that Lee's (1981) result holds for es-

timating ordered Poisson means. *They correct this statement in their 2006 correction note.*

However, there are many cases where Lee–Kelly-like results do hold and we present them below, starting with location parameter cases in this section and scale parameter ones in the next section.

First we go back to the Lee (1988) result for normal means. As seen above, he shows that, for the simple-tree order, the MLE $\hat{\mu}_i$ of μ_i dominates X_i for squared-error loss when $i \neq 1$ and $\nu_1 \leq \nu_i$ for $i \geq 2$. This result has been generalized by Fernández, Rueda and Salvador (1998). They suppose that $X = (X_1, \ldots, X_k)$ has an elliptically symmetric density defined by

$$f(x - \mu) = g\left((x - \mu)' \Sigma^{-1}(x - \mu)\right), \qquad (5.1)$$

with $g(u)$ is non-increasing in u. Then, for Σ known and diagonal, they show that, for the simple-tree order with $\mu_1 \leq \mu_i$, $i = 2, \ldots, k$,

$$P_\mu(|X_i - \mu_i| \leq t_i, i = 1, \ldots, k) \leq P_\mu(|\hat{\mu}_i - \mu_i| \leq t_i, i = 1, \ldots, k) \qquad \text{for all } \mu \in \Omega$$

provided $0 \leq t_i \leq t_1$, $i = 2, \ldots, k$. This result implies that, for $i \neq 1$, $P_\mu(|X_i - \mu_i| \leq t) \leq P_\mu(|\hat{\mu}_i - \mu_i| \leq t)$ for all $t > 0$ and all $\mu \in \Omega$, which implies that, for $i \neq 1$, $\hat{\mu}_i$ universally dominates X_i with respect to the class of loss functions which are non-decreasing in $|d - \mu_i|$. Fernández, Rueda and Salvador (1998) obtain this result from their more general result, which says that, when Ω is such that there does not exist an $i \neq 1$ with $\mu_i \leq \mu_1$, $\hat{\mu}_i$ universally dominates X_i for $i \neq 1$.

Next, let $X_{i,1}, \ldots, X_{i,n_i}$, $i = 1, \ldots, k$, be independent with densities

$$\frac{1}{\nu_i} e^{-(x - \mu_i)/\nu_i} \qquad x > \mu_i,$$

where the ν_i are known. Then, based on the ith sample alone, $X_i = \min(X_{i,1} \ldots, X_{i,n_i})$ is sufficient for μ_i and its density is given by

$$\frac{n_i}{\nu_i} e^{-n_i(x - \mu_i)/\nu_i} \qquad x > \mu_i.$$

Using squared-error loss, the best (i.e., minimum-risk location-equivariant) estimator (MRE) of μ_i based on X_i alone is $X_i - \nu_i/n_i$, $i = 1, \ldots, k$. This is a case where estimators δ_i of μ_i should (see Chapter 2) satisfy the "extra" restriction that, for each $i = 1, \ldots, k$, $\delta_i(X) \leq X_i$ with probability 1 for all $\mu \in \Omega$. As will be seen, not all of the estimators proposed in the literature satisfy this restriction. In cases where it is not satisfied this will be explicitly mentioned.

Vijayasree, Misra and Singh (1995) assume that $\mu_1 \leq \ldots \leq \mu_k$ and consider, for a given $i \in \{1, \ldots, k\}$, estimators of $\theta_1 = \mu_i$. These estimators are of the form $\hat{\mu}_{i,\phi_i}(X) = X_i - \nu_i/n_i + \phi_i(Y_i)$, with $Y_i = (Y_{i,1} \ldots, Y_{i,i-1}, Y_{i,i+1}, \ldots, Y_{i,k})$, $Y_{i,j} = X_i - X_j, i \neq j$. They use squared-error loss and Brewster–Zidek's (1974) first method to obtain explicit dominators of $\hat{\mu}_{i,\phi_i}(X)$. As an example of their results, they show that $X_i - \nu_i/n_i$, as an estimator of μ_i, is dominated by

$$
\delta_i(X) = \begin{cases} \min\left(X_i - \dfrac{\nu_i}{n_i}, \; \hat{\mu}_i(X) - \dfrac{1}{q}\right) & \text{when } i = 1, \ldots, k-1 \\[4mm] \max\left(X_k - \dfrac{\nu_k}{n_k}, \; \hat{\mu}_i(X) - \dfrac{1}{q}\right) & \text{when } i = k, \end{cases} \tag{5.2}
$$

where $q = \sum_{j=1}^{k}(n_j/\nu_j)$ and $\hat{\mu}_i(X) = \min(X_i, \ldots, X_k)$ is the MLE of μ_i. Garren (2000) shows that, for $i = 1$, the estimator (5.2) even universally dominates (with respect to the class of loss functions which are non-decreasing in $|d - \theta_i|$) the MRE on the larger space defined by the simple tree ordering. Further, from the results of Vijayasree, Misra and Singh (1995) it also follows that the MLE $\hat{\mu}_i$ of μ_i is dominated by $\hat{\mu}_i^*(X) = \min(X_i, \ldots, X_k) - 1/q$. Garren (2000) generalizes this result to arbitrary orderings among the parameters as follows. As described in Chapter 8, Section 8.1, an arbitrary ordering among the parameters $\mu_1, \ldots, \mu_k$ can be defined by

$$
\Omega = \{\mu \mid \alpha_{i,j}(\mu_i - \mu_j) \leq 0, 1 \leq i < j \leq k\}, \tag{5.3}
$$

where the $\alpha_{i,j}$ are either 0 or 1 and $\alpha_{i,h} = \alpha_{h,j} = 1$ for some h with $i < h < j$ implies that $\alpha_{i,j} = 1$. Then, for $\mu \in \Omega$, the MLE of μ_i is, for this exponential location problem given by $\hat{\mu}_i(X) = \min\{X_j \mid j \in U_i\}$, where $U_i = \{i\} \cup \{j \mid \alpha_{i,j} = 1\}$ and Garren (2000) proves that, when μ_i is a node, $\hat{\mu}_i(X) - (1/q)$ dominates $\hat{\mu}_i(X)$ for squared-error loss, but does not universally dominate it.

For the particular case where $k = 2$ and $\mu_1 \leq \mu_2$, Pal and Kushary (1992) also obtain dominators of the MREs of μ_1 and μ_2 for squared-error loss. For example, for estimating μ_1 they show that

$$
\delta_1(X) = \begin{cases} X_1 - \dfrac{\nu_1}{n_1} & \text{when } X_1 - X_2 \leq \beta \\[4mm] X_2 - \gamma & \text{when } X_1 - X_2 > \beta, \end{cases}
$$

where β and γ are constants satisfying $\beta \geq (\nu_1/n_1) - \gamma \geq 0$ and

$$
\left(\beta + \gamma - \frac{\nu_1}{n_1}\right)^2 + 2(\beta + \gamma)\left(\frac{\nu_1}{n_1} - \gamma + \frac{\nu_1\nu_2}{n_1\nu_1 + n_2\nu_2}\right) \geq 0,
$$

dominates the MRE of μ_1 based on X_1 alone, i.e. $X_1 - (\nu_1/n_1)$. Vijayasree, Misra and Singh (1995) note that, for $i = 1$ and $k = 2$, (5.2) is a member of this Pal–Kushary class of dominators. Further, Kushary and Cohen's (1989) dominator for squared-error loss of $X_1 - (\nu_1/n_1)$ as an estimator of μ_1 is also a member of this class.

An example of Pal and Kushary's dominators of $X_2 - \nu_2/n_2$ as estimators of μ_2 is given by

$$\begin{cases} X_2 - \dfrac{\nu_2}{n_2} & \text{when } X_2 - X_1 \geq \beta \\ X_1 - \gamma & \text{when } X_2 - X_1 < \beta, \end{cases}$$

where β and γ satisfy $\beta \geq \nu_2/n_2 - \gamma > 0$ and

$$2\left(\left(\beta_o - \frac{\nu_1}{n_1}\right)e^{n_1\beta_o/\nu_1} + \frac{\nu_1}{n_1}\right)\left(\left(\frac{n_1}{\nu_1} + \frac{n_2}{\nu_2}\right)^{-1} + \left(\frac{\nu_1}{n_1} - \frac{\nu_2}{n_2}\right)\right)$$

$$- \beta_o^2 e^{n_1\beta_o/\nu_1} \geq 0,$$

where $\beta_o = \beta - \nu_2/n_2 + \gamma$. This dominator does not satisfy the condition that it is less than X_2 with probability 1 for all $\mu \in \Omega$. But the author's dominator of $X_2 - \nu_2/n_2$ as an estimator of μ_2 given by

$$\begin{cases} X_2 - \dfrac{\nu_2}{n_2} & \text{when } X_2 - X_1 \geq \beta \\ X_2 - \gamma & \text{when } X_2 - X_1 < \beta \end{cases}$$

with $\beta \geq 0$ and γ satisfying

$$2\left(\gamma - \frac{\nu_2}{n_2}\right)\left(\frac{n_1}{\nu_1} + \frac{n_2}{\nu_2}\right)^{-1} - \left(\gamma^2 - \frac{\nu_2^2}{n_2^2}\right) > 0$$

does satisfy this condition when $\gamma \geq 0$. The authors note that the optimal γ is $\gamma_{opt} = (n_1/\nu_1 + n_2/\nu_2)^{-1}$.

A class of so-called mixed estimators of μ_i when $k = 2$ and $\mu_1 \leq \mu_2$ is considered by Misra and Singh (1994). For $i = 1$ these are given by

$$\delta_{1,\alpha}(X) = \begin{cases} X_1 - \dfrac{\nu_1}{n_1} & \text{when } X_1 - \dfrac{\nu_1}{n_1} < X_2 - \dfrac{\nu_2}{n_2} \\[2mm] \alpha\left(X_1 - \dfrac{\nu_1}{n_1}\right) + (1-\alpha)\left(X_2 - \dfrac{\nu_2}{n_2}\right) & \text{when } X_1 - \dfrac{\nu_1}{n_1} \geq X_2 - \dfrac{\nu_2}{n_2}. \end{cases}$$

For squared-error loss these authors show that, when $p = \nu_2 n_1/(\nu_1 n_2) \leq 1$, $\delta_{1,\alpha}$ dominates $\delta_{1,\alpha'}$ when $\alpha^* \leq \alpha < \alpha'$, where $\alpha^* = p^2/(2(p+1))(\leq .25)$. For $p > 1$ they show that $\delta_{1,\alpha}$ dominates $\delta_{1,\alpha'}$ when $\alpha^{**} \leq \alpha < \alpha'$ where

$$\alpha^{**} = 1 - \frac{1 - \dfrac{p^2}{(1+p)^2} e^{(1-p)/p}}{1 + p^2 - \dfrac{2p^3}{1+p} e^{(1-p)/p}} \quad (<1).$$

Given that, for all $\mu \in \Omega$,

$$P_\mu(\delta_{1,\alpha}(X) \le X_1) \begin{cases} = 1 \text{ when } \alpha \le 1 \\ < 1 \text{ when } \alpha > 1, \end{cases}$$

these conditions should be changed to $\alpha^* \le \alpha < \alpha' \le 1$ and $\alpha^{**} \le \alpha < \alpha' \le 1$, respectively.

The authors also have results for estimating μ_2 by the mixed estimator

$$\delta_{2,\beta}(X) = \begin{cases} X_2 - \dfrac{\nu_2}{n_2} & \text{when } X_1 - \dfrac{\nu_1}{n_1} < X_2 - \dfrac{\nu_2}{n_2} \\[2mm] (1-\beta)\left(X_1 - \dfrac{\nu_1}{n_1}\right) + \beta\left(X_2 - \dfrac{\nu_2}{n_2}\right) & \text{when } X_1 - \dfrac{\nu_1}{n_1} \ge X_2 - \dfrac{\nu_2}{n_2}. \end{cases}$$

For this case they show that, when $p \le 1$, δ_{2,β^*} dominates $\delta_{2,\beta}$ when $\beta \ne \beta^*$, where $\beta^* = (2p + 2 - p^2)/(2(p+1))(<1)$; and for $p > 1$ they show that $\delta_{2,\beta}$ dominates $\delta_{2,\beta'}$ when $\beta' < \beta \le \beta^*$ as well as when $\beta^{**} \le \beta < \beta'$, where

$$\beta^{**} = 1 - p^2 \frac{1 + \dfrac{1 - 2p - 2p^2}{(1+p)^2} e^{(1-p)/p}}{1 + p^2 - \dfrac{2p^3}{1+p} e^{(1-p)/p}} \quad (<1).$$

However, for all $\mu \in \Omega$,

$$P_\mu(\delta_{2,\beta}(X) \le X_2) \begin{cases} = 1 \text{ when } \beta \ge 1 \\ < 1 \text{ when } \beta < 1, \end{cases}$$

so, within the class of mixed dominators of the MRE which satisfy the "extra" restriction, the best one is the unrestricted MRE.

Misra and Singh (1994) also give numerical values for the MSE of $\delta_{1,\alpha}$ for several values of α, inclusive α^*. For $\delta_{2,\beta}$ they have MSE values only for $\beta = \beta^*$. These numerical results are presented in Chapter 7, Section 7.2.

The linex loss function has also been considered for the problem of estimating ordered exponential location parameters. Parsian and Sanjari Farsipour (1997) take $k = 2$ and estimate μ_1 with $L(d, \mu_1) = e^{a(d-\mu_1)} - a(d - \mu_1) - 1$ where $a \ne 0$. For the unrestricted case with known ν_1, the best (minimum risk) location-equivariant estimator of μ_1 based on X_1 alone is (see Parsian, Sanjari Farsipour and Nematollahi, 1993) given by $X_1 - \log(n_1/(n_1 - a\nu_1))/a$, provided $a < n_1/\nu_1$. Parsian and Sanjari Farsipour (1997) show that this estimator is improved upon by replacing X_1 by $\min(X_1, X_2)$. These estimators are not scale-equivariant unless $a = a^*/\nu_1$ for some nonzero constant a^*.

Kubokawa and Saleh (1994) consider the more general problem of estimating location parameters when $X_1, \ldots, X_k$ are independent with density $f_i(x - \mu_i)$, $i = 1, \ldots, k$, the μ_i are simply-tree-ordered and μ_1 is the parameter of interest. The densities have strict monotone likelihood ratio in x and the loss function satisfies $L(d, \mu_1) = W(d - \mu_1)$, where $W(y)$ is strictly bowl-shaped and absolutely continuous. Using Kubokawa's (1994b) integral-expression-of-risk method, they give conditions on φ under which estimators of the form $\hat{\mu}_\varphi(X) = X_1 - \varphi(X_2 - X_1, \ldots, X_k - X_1)$ dominate $X_1 - c$, the MRE of θ_1 based on X_1 alone. Examples of their dominators are the generalized Bayes estimators with respect to the prior $d\mu_1 d\mu I(\mu > \mu_1) I(\mu_2 = \ldots = \mu_k = \mu)$ and, for the case where $X_i \sim^{ind} \mathcal{N}(\mu_i, \nu_i^2)$, $i = 1, \ldots, k$ with the ν_i^2 known, the estimator

$$\delta(X) = \min\left(X_1, \frac{\sum_{i=1}^{k} X_i/\nu_i^2}{\sum_{i=1}^{k} 1/\nu_i^2}\right). \tag{5.4}$$

For $k = 2$ but not for $k \geq 3$, this δ is the MLE $\hat{\mu}_1$ of μ_1. As already noted, Lee (1988) shows that, for $k \geq 3$, the MLE of μ_1 does not necessarily dominate X_1 for this normal-mean problem.

An example where the Kubokawa and Saleh (1994) condition of strict monotone likelihood ratio is not satisfied is the exponential location problem of Vijayasree, Misra and Singh (1995). For $k = 2$ and squared-error loss, the Kubokawa–Saleh dominator $X_1 - \varphi_S(X_2 - X_1)$ of $X_1 - \nu_1/n_1$ gives, as is easily seen, $X_1 - \nu_1/n_1$ itself.

The Kubokawa–Saleh (1994) class of estimators is studied by van Eeden and Zidek (2001, 2002, 2004) for the case where $k = 2$ with the $X_i \sim \mathcal{N}(\mu_i, \nu_i^2)$, known ν_i^2 and squared-error loss. They consider the case where $\mu_1 \leq \mu_2$ as well as the case where $|\mu_2 - \mu_1| \leq c$ for a known positive constant c. They compare several estimators of μ_1 of the form $X_1 + \varphi(Z)$, where $Z = X_2 - X_1$. As will be seen in Chapter 7, Section 7.1, they view these estimators as adaptively weighted likelihood estimators. Particular cases are the MLE $\hat{\mu}_1$ with

$$\varphi(Z) = \begin{cases} \dfrac{\min(0, Z)}{1 + \tau} & \text{for } \mu_1 \leq \mu_2 \\[2ex] \dfrac{(Z - c)I(Z > c) + (Z + c)I(Z < -c)}{1 + \tau} & \text{for } |\mu_2 - \mu_1| \leq c, \end{cases} \tag{5.5}$$

the Pitman estimator δ_P (i.e., the first component of the generalized Bayes estimator with respect to the uniform prior on Ω) with

$$
\varphi(Z) = \begin{cases} -\dfrac{\nu_1^2}{\nu} \dfrac{\phi(Z/\nu)}{\Phi(Z/\nu)} & \text{for } \theta_1 \le \theta_2 \\[4mm] \dfrac{\nu_1^2}{\nu} \dfrac{\phi((Z-c)/\nu) - \phi((Z+c)/\nu)}{\Phi((Z+c)/\nu) - \Phi((Z-c)/\nu)} & \text{for } |\theta_2 - \theta_1| \le c \end{cases} \tag{5.6}
$$

and the estimator δ_{WLE} with

$$
\varphi(Z) = \begin{cases} \dfrac{Z\nu_1^2}{\nu^2 + (\max(0, Z))^2} & \text{for } \mu_1 \le \mu_2 \\[4mm] \dfrac{Z\nu_1^2}{\nu^2 + \min(Z^2, c^2)} & \text{for } |\mu_2 - \mu_1| \le c, \end{cases} \tag{5.7}
$$

where $\tau = \nu_2^2/\nu_1^2$ and $\nu^2 = \nu_1^2 + \nu_2^2$. They study and compare these estimators analytically as well as numerically. They show, e.g., that, in both cases, $\hat{\mu}_1$ and δ_P dominate X_1. Further (again in both cases), $\hat{\mu}_1$ and δ_{WLE} are, among estimators based on (X_1, X_2), inadmissible, while δ_P is admissible in this class of estimators. Dominators for some of these inadmissible estimators as well as minimax estimators and (references to) proofs of the admissibility of δ_P are, for both cases, presented in Section 5.5. The authors' numerical results concerning the MSEs of their estimators as well as their robustness with respect to misspecification of Ω, are discussed in Chapter 7, Section 7.2.

Remark 5.3. For the case where $X_i \sim^{ind} \mathcal{N}(\mu_i, \nu_i^2)$, $i = 1, \ldots, k$, with simply-tree-ordered μ_i, known ν_i^2's and squared-error loss, the above-given Kubokawa–Saleh (1994) dominator (5.4) of X_1 is identical to the one Brewster and Zidek (1974) obtain by their first method (see Brewster and Zidek, 1974, formula (2.2.2)). Also, for the same problem but with $k = 2$, the generalized Bayes estimator of Kubokawa and Saleh is the same as the one Brewster and Zidek obtain by their third method – both are generalized Bayes with respect to a uniform prior on Ω. For $k > 2$, these generalized Bayes estimators are not the same. Kubokawa and Saleh do not mention this overlap of their results with those of Brewster and Zidek other than (see Kubokawa and Saleh, 1994, p. 41, lines -9 to -7) the fact that Brewster and Zidek (and others) have demonstrated that the ordinary estimator is improved on by using the restriction. Nor do Kubokawa and Saleh explore whether, for problems other than the simply-tree-ordered normal-mean problem, some of their results could have been obtained by using one of the Brewster–Zidek methods.

And van Eeden and Zidek (2002), for the ordered normal-mean problem with $k = 2$, apparently forgot that for two of their estimators, namely, the MLE and the Pitman estimator, Brewster–Zidek (1974) already proved that they dominate X_1.

Results for general location problems can also be found in Hwang and Peddada (1994). One of their results is concerned with the i-th component,

$\hat{\mu}_i^{so}(X)$, of the MLE of μ under the simple-order restriction. They assume that $X = (X_1, \ldots, X_k)$ has an elliptically symmetric density defined by (5.1) and show that, when Σ is known and diagonal, this estimator universally dominates X_i (with respect to the class of loss functions which are nondecreasing in $|d - \mu_i|$), for any parameter space Ω defined by inequalities among the components of μ when μ_i is a node. Clearly, the Lee–Kelly result for the normal-mean problem with $\Omega = \{\mu \mid \mu_1 \leq \ldots \leq \mu_k\}$ and known variances is a special case of this Hwang–Peddada result. It also implies, for the density (5.1) with known diagonal Σ, that $\hat{\mu}_1^{SO}(X) = \min_{t \geq 1}(\sum_{i=1}^{t} X_i/\nu_i^2 / \sum_{i=1}^{t} 1/\nu_i^2)$ universally dominates X_1 when the μ_i are tree-ordered, but I do not know whether $\hat{\mu}_1^{so}$ and (5.4), the Brewster–Zidek (1974) (Kubokawa–Saleh, 1994) dominator of X_1 for the simple tree-ordered normal-mean problem, are comparable for squared-error loss. Hwang and Peddada (1994) also have results for the case where Σ is not diagonal.

Remark 5.4. Hwang and Peddada (1994) state their results in terms of the isotonic regression estimator of μ with weights $w_i > 0$, $i = 1, \ldots, k$, with respect to the given ordering of the μ_i. This estimator minimizes (see Chapter 8, Section 8.2), for $\mu \in \Omega$, $\sum_{i=1}^{k}(X_i - \mu_i)^2 w_i$ and is thus the MLE of μ when X has density (5.1) with diagonal known Σ and the w_i are the diagonal elements of Σ^{-1}.

Iliopoulos (2000) uses the Kubokawa and Saleh (1994) results to obtain a dominator for the MRE, $X_2 - c$, based on X_2 alone of the middle one of three simply-ordered location parameters. Under the distributional assumptions of Kubokawa and Saleh (1994) and for a strictly convex loss function, he first finds a dominator $\mu_\phi(X_2, X_3) = X_2 - \phi(X_3 - X_2)$ for $X_2 - c$ when $\mu_2 \leq \mu_3$. Then he finds ψ such that $\delta_\psi(X_1, X_2) = X_2 - \psi(X_1 - X_2)$ dominates $X_2 - c$ when $\mu_1 \leq \mu_2$. He then shows that $X_2 - \psi(X_1 - X_2) - \phi(X_3 - X_2) + c$ dominates $X_2 - c$ when the $\mu_1 \leq \mu_2 \leq \mu_3$. For the normal-mean case Ilioupoulos also gives numerical values for the percent risk improvement of his estimator relative to the unrestricted MRE of μ_2. Some of these are presented in Chapter 7, Section 7.2.

Finally, another case of restrictions other than inequalities among the parameters. Let $X_i \sim^{ind} \mathcal{N}(\mu_i, 1)$, $i = 1, \ldots, k$ and $\Omega = \{\mu \mid \sum_{i=1}^{k} \mu_i^2 \leq m^2\}$ for some known $m > 0$ and let the loss be squared-error. Then the MLE of μ_i is given by

$$\hat{\mu}_i = \begin{cases} X_i & \text{when } |X| \leq m \\ \\ mX_i/|X| & \text{when } |X| > m, \end{cases}$$

where $|X|^2 = \sum_{i=1}^{k} X_i^2$. Hwang and Peddada (1993) show, as a special case of a more general result, that $\hat{\mu}_1$ dominates X_1 when $m \leq 1$, but that for $m > 1$ and large enough k, $\hat{\mu}_1$ fails to dominate X_1.

5.2 Scale parameter estimation with known ν

In this section we present Lee–Kelly-like results when scale parameters are to be estimated and $\nu = (\nu_1, \ldots, \nu_k)$ is known.

Misra and Dhariyal (1995) consider the case where the $X_{i,j}$ are $\mathcal{U}(0, \mu_i)$ with $0 < \mu_1 \leq \ldots \leq \mu_k$ and scale-invariant squared-error loss. Let, for $i = 1, \ldots, k$, $Y_i = \max_{1 \leq j \leq n_i} X_{i,j}$, $Y_i^* = \max(Y_1, \ldots, Y_i)$ and $Y = (Y_1, \ldots, Y_k)$. Then Y_i (the unrestricted MLE of μ_i) is sufficient for μ_i, $i = 1, \ldots, k$ and the best (i.e., minimum-risk scale-equivariant estimator (MRE)) of μ_i based on Y_i alone is $(n_i+2)Y_i/(n_i+1)$. This is, again, a case where the estimators $\delta_i(Y)$ of μ_i should satisfy an extra restriction, namely, that, for each $i = 1, \ldots, k$, $\delta_i(Y) \geq Y_i$ with probability 1 for all $\mu \in \Omega$. The authors use Brewster and Zidek's (1974) first method to show that $(n_i + 2)Y_i/(n_i + 1)$ is inadmissible as an estimator of μ_i and is dominated by δ_i, where

$$\delta_1(Y) = \min \left\{ \frac{n_1 + 2}{n_1 + 1} Y_1, \frac{n + 2}{n + 1} Y_k^* \right\}$$

$$\delta_i(Y) = \max \left\{ \frac{n_i + 2}{n_i + 1} Y_i, \frac{n + 2}{n + 1} Y_i^* \right\}, \qquad i = 2, \ldots, k,$$

(5.8)

with $n = \sum_{j=1}^{k} n_j$. More generally, they give dominators for estimators of the form $\hat{\mu}_{i,\phi_i}(Y) = Y_1 \phi_i(Z)$, where $Z = (Y_2/Y_1, \ldots, Y_k/Y_1)$. But Lillo and Martín (2000) show, for $k = 2$, that δ_2 is inadmissible for squared-error loss as an estimator of μ_2 and dominated by $\delta_2^*(Y)$ given by

$$\delta_2^*(Y) = \begin{cases} \delta_2(Y) & \text{when } Y_2 \geq \dfrac{n + 2}{n + 1} \dfrac{n_2 + 1}{n_2 + 2} Y_1 \\[2ex] \dfrac{n + 2}{n + 1} \dfrac{Y_1^2}{Y_2} & \text{otherwise.} \end{cases}$$

It is not difficult to see that each of the estimators δ_i, $i = 1, \ldots, k$, as well as the estimator δ_2^* satisfy the extra restrictions with probability 1 for all $\mu \in \Omega$.

For this ordered uniform-scale-parameter problem with $k = 2$ and squared-error loss, Joorel and Hooda (2002) show that $Y_2^* = \max(Y_1, Y_2)$, the MLE of μ_2, dominates Y_2, the unrestricted MLE of μ_2, while, for estimating μ_1, they claim that the MLE is dominated by Y_1 not realizing that Y_1 is the MLE and not (as they say) $\min(Y_1, Y_2)$. These authors also consider estimators of μ_1 of the form $\delta(Y_1, Y_2) = cY_1 I(Y_1 \leq Y_2) + dY_1 I(Y_1 > Y_2)$ and show that such an estimator dominates the unrestricted MLE (and thus the MLE) when

$$1 \leq c \leq \frac{n + 3}{n + 1} \text{ and } 4(n + 1)/(2n + 1) - c \leq d \leq c$$

or

$$1 \le c \le \frac{n+3}{n+1} \text{ and } c \le d \le \frac{4(n+1)}{2n+1} - c.$$

They give similar results for estimating μ_2.

Joorel and Hooda (2002) also give what they call "optimal estimators" of μ_1 and μ_2. For estimating μ_1 these estimators are of the form $c\min(Y_1, Y_2)$ and they find that the MSE of this estimator is minimized for

$$c = \frac{n+2}{2n+1} \left(\frac{2n + 1 - (\mu_1/\mu_2)^n}{n + 1 - (\mu_1/\mu_2)^n} \right). \tag{5.9}$$

But this of course does not help, because this c depends on the μ's. In their 2005 (submitted) correction note they add the condition that μ_1/μ_2 is known and then of course the c in (5.9) gives an optimal estimator of μ_1 among those of the form $c\min(Y_1, Y_2)$. But there is another problem with this estimator of μ_1: it does not satisfy the "extra" restriction that $c\min(Y_1, Y_2) > Y_1$ with probability 1 for all $\mu \in \Omega$. Similar results for optimally estimating μ_2 by $c\max(Y_1, Y_2)$ (again adding the condition that μ_1/μ_2 is known) give

$$c = \frac{n+2}{2n+1} \left(\frac{2n + 1 + (\mu_1/\mu_2)^{n+1}}{n + 1 + (\mu_1/\mu_2)^{n+2}} \right).$$

This estimator does satisfy the condition of being larger than Y_2 with probability 1 for all $\mu \in \Omega$. But note that, when $\mu_1/\mu_2 = r \in (0, 1]$ is known, the problem is not anymore a problem in restricted-parameter-space estimation. In fact, there is only one unknown parameter, μ_2 say, and $X_{1,1}/r, \ldots, X_{1,n}/r, X_{2,1}, \ldots, X_{2,n}$ are independent $\mathcal{U}(0, \mu_2)$ so that $(2n + 2)\max(Y_1/r, Y_2)/(2n + 1)$ has minimum-risk among scale-invariant estimators of μ_2. Of course, the Joorel–Hooda estimator $c\max(Y_1, Y_2)$ is also scale-invariant, but has a larger MSE because it minimizes the MSE over a smaller class of estimators.

Remark 5.5. In their 2005 (submitted) correction note, Joorel and Hooda correct most of the misprints in their 2002 paper. However, not all of the mistakes in their paper are corrected in that note. Their reasoning in their Section 3 that there does not exist a function of (Y_1, Y_2) which is unbiased for estimating μ_1 and μ_2 is incorrect. They do not mention this in their correction note.

More results for this uniform scale parameter problem can be found in Fernández, Rueda and Salvador (1997). Their parameter space is defined by an arbitrary ordering among the μ_i which can (see (5.3)) be described by

$$\Omega = \{\mu \mid \alpha_{i,j}(\mu_i - \mu_j) \le 0, 1 \le i < j \le k\},$$

where the $\alpha_{i,j}$ are either 0 or 1 and $\alpha_{i,h} = \alpha_{h,j} = 1$ for some h with $i < h < j$ implies that $\alpha_{i,j} = 1$. Let $L_i = \{i\} \cup \{j < i \mid \alpha_{j,i} = 1\}$, then (see Chapter 8, Section 8.2), the MLE of μ_i is $\hat{\mu}_i = \max_{j \in L_i} Y_j$. Fernández, Rueda and Salvador then show that, when $L_i \neq i$, $\hat{\mu}_i$ universally dominates Y_i (the unrestricted MLE of μ_i) with respect to the class of loss functions which are nondecreasing in $|d-\mu_i|$. Note that, when $i = 1$, $L_i = \{i\}$ and $\max_{j \in L_i} Y_l = Y_1$ so that for $i = 1$, $\hat{\mu}_1 = Y_1$. For the simple tree order, these authors also show that, depending on k, c and the n_i, $c'\hat{\mu}$ and $c'Y$ can be noncomparable as estimators of $c'\mu$.

Another scale-parameter problem is the case where the $X_{i,j}$, $j = 1, \ldots, n_i$, have density

$$f(x; \mu_i) = \frac{1}{\mu_i} e^{-(x/\mu_i)}, \qquad x > 0, \mu_i > 0, i = 1, \ldots, k.$$

The sufficient statistic for μ_i is $X_i = \sum_{j=1}^{n_i} X_{i,j}$, its density is given by

$$\frac{1}{\Gamma(n_i)\,\mu_i^{n_i}}\, x^{n_i-1} e^{-x/\mu_i}, \qquad x > 0, \tag{5.10}$$

the unrestricted MLE of μ_i is $\bar{X}_i = X_i/n_i$ and its MRE for scale-invariant squared-error loss based on X_i alone is $X_i/(n_i + 1)$. For the simple order, Vijayasree, Misra and Singh (1995) consider estimators of μ_i of the form $\hat{\mu}_{i,\phi_i}(X) = X_i\phi_i(Z_i)$, where, with $Z_{j,i} = X_j/X_i$, $j \neq i$, $i = 1, \ldots, k$, $Z_i = (Z_{1,i}, \ldots, Z_{i-1,i}, Z_{i+1;i}, \ldots, Z_{k,i})$. In the same way as for their location parameter results, they use Brewster–Zidek's (1974) first method to obtain explicit dominators of the of these estimators. They show, e.g., that

$$\delta_i(X) = \begin{cases} \min\left(\dfrac{X_1}{n_1+1}, \dfrac{\sum_{j=1}^{k} X_j}{1+p}\right) & \text{when } i = 1 \\[3ex] \max\left(\dfrac{X_i}{n_i+1}, \dfrac{\sum_{j=1}^{i} X_i}{1+p}\right) & \text{when } i \geq 2, \end{cases}$$

where $p = \sum_{j=1}^{k} n_j$, dominates $X_i/(n_i + 1)$ as an estimator of μ_i. These authors also give the following dominator of the MLE $\hat{\mu}_i$ of μ_i, $i = 1, \ldots, k-1$,

$$\delta_i^*(X) = \begin{cases} \min\left(\hat{\mu}_1, \dfrac{X_1}{1+p}(1 + \sum_{j=2}^{k} Z_{j,1})\right) & \text{when } i = 1 \\[3ex] \max\left(\hat{\mu}_i, \dfrac{X_i}{1+p}(1 + \sum_{j=1}^{i-1} Z_{j,i})\right) & \text{when } i = 2, \ldots, k-1. \end{cases}$$

Kaur and Singh (1991) consider the special case where $k = 2$ and $n_1 = n_2 = n$. They show that, for estimating μ_i, $i = 1, 2$, the MLE of μ_1 dominates X_1/n and the MLE of μ_2 dominates X_2/n when $n \geq 2$. Further, Vijayasree and

Singh (1993) show, also for $k = 2$, that X_1/n_1, as an estimator of μ_1, is dominated by the so-called mixed estimator $\delta_{1,\alpha}(X) = \min(X_1/n_1, \alpha(X_1/n_1) + (1-\alpha)(X_2/n_2))$ when $\alpha_1 = n_1/(n_1+n_2+1) \leq \alpha < 1$. This mixed estimator is the MLE of μ_1 when $\alpha = n_1/(n_1+n_2)$ and it equals X_1/n_1 when $\alpha = 0$. This result implies that the MLE dominates X_1/n_1. For estimating μ_2, the authors use the mixed estimator $\delta_{2,\alpha}(X) = \max(X_2/n_2, \alpha X_1/n_1 + (1-\alpha)X_2/n_2)$ and show that, for $0 < \alpha \leq n_1(2n_1+n_2)/((n_1+n_2)(n_1+n_2+1)) = \alpha_2$, $\delta_{2,\alpha}$ dominates X_2/n_2. This mixed estimator is the MLE of μ_2 when $\alpha = n_1/(n_1+n_2)$ implying that the MLE dominates X_2/n_2. They further show that, for $\alpha^* = (1/2) - (1/2)^{2n} \binom{2n-1}{n}$, δ_{2,α^*} dominates $\delta_{2,\alpha}$ for $\alpha \in [0, \alpha^*)$. These results imply, of course, that $(\delta_{1,\alpha}(X), \delta_{2,\alpha}(X))$ as an estimators of (μ_1, μ_2) dominates $(X_1/n_1, X_2/n_2)$ for the sum of the squared-error losses. But Vijayasree and Singh (1991) (a paper Vijayasree and Singh (1993) do not refer to) show that this domination of $(X_1/n_1, X_2/n_2)$ holds for all $\alpha \in (0, 1)$. Vijayasree and Singh (1993) also give numerical comparisons of their estimators. Some of those results are presented in Chapter 7, Section 7.2.

Remark 5.6. Using the first method of Brewster and Zidek (1974) does not give a dominator for the MLE $\hat\mu_k$ of μ_k and it seems to be unknown whether it is admissible. Another unknown seems to be whether $\hat\mu_k$ and X_k/n_k are comparable, except (as noted above from the results of Vijayasree and Singh (1993) and Kaur and Singh (1991)) for the case where $k = 2$.

More results for $k = 2$ for this gamma-scale problem, still with scale-invariant squared-error loss, can be found in Misra, Choudhary, Dhariyal and Kundu (2002). They note that Vijayasree, Misra and Singh's (1995) dominators being "non-smooth" might well be inadmissible and they, therefore, look for "smooth" dominators. The estimators they dominate are the MREs of μ_1 and μ_2. For estimating μ_1 (with similar results for estimating μ_2) they start with a class of non-smooth estimators of the form

$$\phi_1(c, r, T) = \begin{cases} \dfrac{X_1}{n_1+1} & \text{if } Z \geq r \\ cX_1 & \text{if } Z < r, \end{cases}$$

where $T = (X_1, X_2)$, $Z = X_2/X_1$ and r and c are fixed positive constants. Note that, for all $r > 0$, $\phi_1(1/(n_1+1), r, T) = X_1/(n_1+1)$, the MRE of μ_1. The authors then study the risk function of $\phi_1(c, r, T)$ as a function of c for fixed r and find that, for each $r > 0$, $\phi_1(c(r), r, T)$ dominates $X_1/(n_1+1)$, where

$$c(r) = \frac{1}{n_1+1} \left(1 - \frac{\displaystyle\int_0^1 \frac{x^{n_1+n_2}}{(1+rx)^{n_1+n_2+2}} dx}{\displaystyle\int_0^1 \frac{x^{n_2-1}}{(1+rx)^{n_1+n_2+2}} dx} \right).$$

The authors then consider, for $0 < r' < r$, the class of (again non-smooth) estimators

$$\phi_2(c, r', r, T) = \begin{cases} \dfrac{X_1}{n_1 + 1} & \text{if } Z \geq r \\[2mm] c(r)X_1 & \text{if } r' \leq Z < r \\[2mm] cX_1 & \text{if } Z < r' \end{cases}$$

and show that $\phi_1(c(r), r, T)$ is dominated by $\phi_2(c(r'), r', r, T)$. Then, using Brewster and Zidek's (1974) third method, the authors select, for each $l = 1, 2, \ldots$, a finite partition of $[0, \infty)$ represented by $0 = r_{l,0} < r_{l,1} < \ldots, r_{l,N_l} = \infty$ and a corresponding estimator defined by

$$\phi_l(T) = \begin{cases} \dfrac{X_1}{n_1 + 1} & \text{when } Z \geq r_{l,N_l-1} \\[2mm] c(r_{l,N_l-1})X_1 & \text{when } r_{l,N_l-2} \leq Z < r_{l,N_l-1} \\[2mm] \vdots \\[2mm] c(r_{l,2})X_1 & \text{when } r_{l,1} \leq Z < r_{l,2} \\[2mm] c(r_{l,1})X_1 & \text{when } Z < r_{l,1}. \end{cases}$$

The authors then show that this sequence of estimators converges pointwise to $X_1 c(S)$ provided

$$\max_j |r_{l,j} - r_{l,j-1}| \to 0 \text{ and } r_{l,N_1-1} \to \infty \text{ as } l \to \infty.$$

In the final step of their proof they show that $X_1 c(Z)$ dominates $X_1/(n_1+1)$. As a further property, the authors show that this dominator $X_1 c(Z)$ is the first component of the Bayes estimator of (μ_1, μ_2) with respect to the uniform prior for $(\log \mu_1, \log \mu_2)$ on $\{\mu \mid \mu_1 \leq \mu_2\}$, i.e., the Pitman estimator. Misra, Choudhary, Dhariyal and Kundu (2002) also show that their dominator $X_1 c(Z)$ of $X_1/(n_1+1)$ can be written in the form

$$\frac{X_1}{n_1+1} \frac{I_{n_2,n_1+1}\left(\dfrac{Z}{Z+1}\right)}{I_{n_2,n_1+2}\left(\dfrac{Z}{Z+1}\right)}, \tag{5.11}$$

where, for $\beta > 0$, $\gamma > 0$ and $0 < x < 1$,

$$I_{\beta,\gamma}(x) = \frac{\Gamma(\beta+\gamma)}{\Gamma(\beta)\Gamma(\gamma)} \int_0^x y^{\beta-1}(1-y)^{\gamma-1} dy.$$

Then using the fact that, for positive γ and β,

$$\int_0^x \frac{y^{\beta-1}}{(1+y)^{\gamma+\beta}} dy = \frac{\Gamma(\beta)\Gamma(\gamma)}{\Gamma(\gamma+\beta)} I_{\beta,\gamma}\left(\frac{x}{x+1}\right)$$

it follows that (5.11) can also be written in the form

$$\frac{X_1}{n_1 + n_2 + 1} \frac{\int_0^Z \frac{y^{n_2-1}}{(1+y)^{n_1+n_2+1}} dy}{\int_0^Z \frac{y^{n_2-1}}{(1+y)^{n_1+n_2+2}} dy},$$

which is the dominator of $X_1/(n_1+1)$ of Kubokawa and Saleh (1994) for the same problem. This Kubokawa–Saleh dominator can be found on their page 50, line -6, where it is given for the (equivalent) case of scale parameters of χ^2 distributions. So, the Misra, Choudhary, Dhariyal and Kundu (2002) dominator is not new, but the authors do not give any credit to Kubokawa–Saleh for this result. What is new in the Misra–Choudhari–Dhariyal–Kundu paper is that they show that Brewster–Zidek's (1974) third method applied to two ordered gamma-scale parameters and scale-invariant squared-error loss gives the generalized Bayes estimator with respect to the uniform distribution for $(\log \mu_1, \log \mu_2)$ for $\mu \in \Omega$, i.e. the Pitman estimator.

Misra, Choudhary, Dhariyal and Kundu's (2002) dominator of $X_2/(n_2+1)$ is given by

$$\frac{X_2}{n_2 + 1} \frac{I_{n_2+1,n_1}\left(\frac{Z}{Z+1}\right)}{I_{n_2+2,n_1}\left(\frac{Z}{Z+1}\right)}. \tag{5.12}$$

Misra, Choudhary, Dhariyal and Kundu (2002) do not say anything about the (in)admissibility of their dominators, but they present Monte-Carlo comparisons of their dominator with, for estimating μ_1, the unrestricted MRE $X_1/(n_1+1)$, the (restricted) MLE and the Vijayasree, Misra and Singh (1995) dominator $\min(X_1/(n_1+1), (X_1+X_2)/(n_1+n_2+1))$ for several values of (n_1, n_2). These numerical results are presented and discussed in Chapter 7, Section 7.2.

Of course, the above gamma-scale results apply to the problem of estimating ordered normal variances with known means as well as with estimated means when $k \geq 2$ samples are available. Kushary and Cohen (1989) and Kourouklis (2000) have results for the problem in this normal-variance form when $k = 2$, but Kourouklis (2000) does not refer to Kushary and Cohen (1989), nor does he refer to Vijayasree, Misra and Singh (1995).

Results for a very general scale problem can be found in Kubokawa and Saleh (1994). They have $X_1, \ldots, X_k$ independent with densities $f_i(x/\mu_i)/\mu_i$, $i = 1, \ldots, k$, $x > 0$, with strict monotone likelihood ratio in x. The μ_i are simply-tree-ordered, μ_1 is to be estimated and the loss function $L(d, \mu_1) = W(d/\mu_1)$, where W is bowl-shaped and absolutely continuous with $W(1) = 0$. They look at a class of estimators of the form $\delta(X) = X_1\phi(X_1/X_1,\ldots,X_k/X_1)$ and give sufficient conditions on ϕ for $\delta(X)$ to dominate the MRE of μ_1 based on X_1 alone. In one of their examples $\delta(X)$ is the generalized Bayes estimator of μ_1 with respect to the prior $(d\mu_1/\mu_1)(d\mu/\mu)I(\mu > \mu_1)I(\mu_1 = \ldots = \mu_k = \mu)$. For the particular case where $k = 2$, some of the Kubokawa–Saleh (1994) results can also be found in Kubokawa (1994a). This last paper also contains results for the estimation of μ_2. And, as already mentioned above, Misra, Choudhary, Dhariyal and Kundu (2002) show that, for $k = 2$, this generalized Bayes estimator can be obtained by Brewster and Zidek's (1974) third method.

More results on scale-parameter estimation can be found in Hwang and Peddada (1994). They consider, for squared-error loss, the estimation of μ_1 and of μ_k when X_i, $i = 1,\ldots,k$ are independent with density (5.10). For the estimation of μ_i they use the i-th component $\hat{\mu}_i^{so}$ of the isotonic regression $\hat{\mu}^{so}$ of μ with respect to the ordering $\mu_1 \leq \ldots \leq \mu_k$ and with weights n_i. This isotonic regression estimator minimizes (see Chapter 8, Section 8.2) $\sum_{i=1}^k (X_i/n_i - \mu_i)^2 n_i$ for $\mu \in \Omega$. The estimator $\hat{\mu}^{so}$ is, for this gamma-scale case, also its MLE for $\mu \in \Omega$. They show that $\hat{\mu}_1^{so}$ universally dominates the unrestricted MLE based on X_1 alone, i.e., X_1/n_1, when Ω is such that $\mu_i \geq \mu_1$ for all $i = 1,\ldots,k$. Further, they claim (in their Theorem 4.6, part ii)) that for general scale families $\hat{\mu}_k^{so}$ does not universally dominate X_k/n_k when μ is such that $\mu_i \leq \mu_k$ for all $i = 1,\ldots,k$. However, as Garren (2000) remarks, the Hwang–Peddada results are based on the assumption that the support of the X_i as well as of the unrestricted estimator are either unbounded or else totally bounded. Further, it can easily be seen from Hwang and Peddada's proof of their Theorem 4.6, part ii), that their proof does not work for the case where independent $X_{i,j}$, $j = 1,\ldots,n_i$, are $\mathcal{U}(0,\mu_i)$, $i = 1,\ldots,k$. For this case Fernández, Rueda and Salvador (1997) show (as already noted) that the MLE of μ_k universally dominates the unrestricted MLE and these authors remark that, therefore, the Hwang–Peddada result for general scale families is wrong. But this reasoning does not work, because Hwang and Peddada do not make any claims about the MLE. They only consider isotonic regression estimators, which for the gamma-scale case are the MLE, but not for the uniform-scale case. A similar remark holds for the results in Hwang–Peddada's Theorem 4.7 concerning the tree-ordered case.

To compare some of the above results, take the particular case where the X_i are independent with density (5.10), $k = 2$, $n_1 = n_2 = n$ and the loss is squared error. With $Z = X_2/X_1$, the dominators of $X_1/(n+1)$ as an estimator of μ_1 become

$$\delta_1(X) = \frac{X_1}{n+1} \min\left(1, \frac{2}{3}(Z+1)\right) \text{ for } n \geq 3 \text{ by Kushary and Cohen (1989)};$$

$$\delta_2(X) = \frac{X_1}{n+1} \min\left(1, \frac{n+1}{2n-1}(Z+1)\right) \qquad \text{by Kourouklis (2000)};$$

$$\delta_3(X) = \frac{X_1}{n+1} \min\left(1, \frac{n+1}{2n+1}(Z+1)\right) \quad \left\{ \begin{array}{l} \text{by Kubokawa and Saleh (1994),} \\ \text{by Kubokawa (1994a)} \\ \text{and} \\ \text{by Vijayasree, Misra} \\ \text{and Singh (1995);} \end{array} \right.$$

and

$$\delta_4(X) = \frac{X_1}{n+1}\left(1 - \frac{\displaystyle\int_0^1 \frac{x^{2n}}{(1+Zx)^{2n+2}}dx}{\displaystyle\int_0^1 \frac{x^{n-1}}{(1+Zx)^{2n+2}}dx}\right) \quad \left\{ \begin{array}{l} \text{by Kubokawa and Saleh} \\ \text{(1994) and} \\ \text{by Misra, Choudhary,} \\ \text{Dhariyal and Kundu} \\ \text{(2002).} \end{array} \right.$$

Now, from Theorem 2.1(b) of Vijayasree, Misra and Singh (1995), we know that, when $P_\theta(X_1\phi(X_2/X_1) > (X_1 + X_2)/(2n + 1)) > 0$ for all $\theta_1 \leq \theta_2$, $X_1\phi(X_2/X_1)$ is dominated by $\min(X_1\phi(X_2/X_1),(X_1 + X_2)/(2n + 1))$. Using this result shows that δ_3 dominates both δ_1 and δ_2. Further, by Kourouklis (2000), $\min(X_1\phi(X_2/X_1),(X_1 + X_2)/(2n - 1))$ is a dominator of $X_1\phi(X_2/X_1)$ when $P_\theta(X_1\phi(X_2/X_1) \neq (X_1+X_2)/(2n+1)) > 0$ for all $\theta_1 \leq \theta_2$. From this result we see that δ_2 dominates δ_1 when $n \geq 6$, while $\delta_1 \equiv \delta_2$ when $n = 5$. (Note that Kourouklis (2000) claims that δ_2 dominates δ_1 for all $n \geq 3$). For dominators of X_1/n for this gamma-distribution setting, the results of Kaur and Singh (1991) as well as those of Hwang and Peddada (1994) give $\min(X_1/n,(X_1 + X_2)/(2n))$, which, by Vijayasree, Misra and Singh (1995), is dominated by $\min(X_1/n,(X_1 + X_2)/(2n + 1))$. Whether δ_4 and/or δ_5 dominate or are dominated by one or more of δ_1, δ_2 and δ_3 seems to be unknown.

Still for the density (5.10), Chang and Shinozaki (2002) give, for $k = 2$, conditions on c_1 and c_2 for $\sum_{i=1}^2 c_i\hat{\mu}_i$ to dominate $\sum_{i=1}^2 c_iX_i/n_i$, as well as conditions for $\sum_{i=1}^2 c_i\hat{\mu}_i$ to dominate $\sum_{i=1}^2 c_iX_i/(n_i + 1)$ as estimators of $\sum_{i=1}^2 c_i\mu_i$. The two special cases $c_1 = 0$ and $c_2 = 0$ each give results which overlap with some of the results of Kaur and Singh (1991).

Remark 5.7. Both Kushary and Cohen (1989) and Kourouklis (2000) assume that the shape parameters in their gamma distributions are integers when multiplied by 2. Further, Kaur and Singh (1991), Vijayasree and Singh (1993) and Vijayasree, Misra and Singh (1995) assume that these parameters are integers. However, only Kaur and Singh (1991) and Vijayasree and Singh (1993) make use of this assumption in their proofs.

5.3 Unknown ν

We now look at results for the case where $\nu = (\nu_1, \ldots, \nu_k)$ is unknown and start with the following problem. Let $X_{i,1}, \ldots, X_{i,n_i}$, $i = 1, \ldots, k$ be independent with $X_{i,j} \sim \mathcal{N}(\mu, \nu_i^2)$. The parameter to be estimated is μ and for this problem Graybill and Deal (1959) propose the use of

$$\hat{\mu}_{GD}(X) = \frac{\sum_{i=1}^{k}(n_i \bar{X}_i)/S_i^2}{\sum_{i=1}^{k} n_i/S_i^2},$$

where $n_i \bar{X}_i = \sum_{j=1}^{n_i} X_{i,j}$ and the $S_1^2, \ldots, S_k^2$ are independent and independent of the $X_{i,j}$ with, for $i = 1, \ldots, k$, $m_i S_i^2/\nu_i^2 \sim \chi_{m_i}^2$ for some $m_i \geq 1$. Note that $\hat{\mu}_{GD}(X)$ can also be written in the form

$$\hat{\mu}_{GD}(X) = \bar{X}_1 + \sum_{i=2}^{k}(\bar{X}_i - \bar{X}_1)\phi_i = \bar{X}_1 \left(1 - \sum_{i=2}^{k} \phi_i\right) + \sum_{i=2}^{k} \bar{X}_i \phi_i,$$

where $\phi_i = n_i/S_i^2/(\sum_{j=1}^{k} n_j/S_j^2)$. Of course, one can take $(n_i - 1)S_i^2 = \sum_{j=1}^{n_i}(X_{i,j} - \bar{X}_i)^2$. One then needs $n_i \geq 2$.

Now suppose that we know that, for some $k_1 \in \{2, \ldots, k\}$, $0 < \nu_1^2 \leq \nu_i^2$ for $i = 2, \ldots, k_1$ and $\nu_i^2 > 0$ for $i = k_1 + 1, \ldots, k$. Then Sinha (1979) shows that, when $n_i = n$, $i = 1, \ldots, k$, $\hat{\mu}_{GD}$ is, on this restricted parameter space, inadmissible as an estimator of μ for the loss function $L(d, \mu) = W(d - \mu)$ with $W(y) = W(-y)$ for all y, $W(y)$ strictly increasing and $\int_0^\infty W(cy)\phi(y)dy < \infty$ for all $c > 0$ and he gives the following dominator

$$\hat{\mu}_S(X) = \bar{X}_1 \left(1 - \sum_{i=k_1+1}^{k} \phi_i - \sum_{i=2}^{k_1} \phi_i^*\right) + \sum_{i=k_1+1}^{k} \bar{X}_i \phi_i + \sum_{i=2}^{k_1} \bar{X}_i \phi_i^*,$$

where $\phi_i^* = \min(\phi_i, 1/2)$.

For $k = 2$ this dominator becomes

$$\begin{cases} \hat{\mu}_{GD}(X) & \text{when } S_1^2 \leq S_2^2 \\[2mm] \dfrac{\bar{X}_1 + \bar{X}_2}{2} & \text{when } S_1^2 > S_2^2. \end{cases}$$

For the particular case where $k = 2$, Elfessi and Pal (1992) show that the Graybill–Deal estimator is universally inadmissible and give two dominators, one for equal sample sizes and one for possibly unequal ones. Their estimator for equal sample sizes is

$$\begin{cases} \hat{\mu}_{GD}(X) & \text{when } S_1^2 \leq S_2^2 \\[2mm] \dfrac{S_1^2 \bar{X}_1 + S_2^2 \bar{X}_2}{S_1^2 + S_2^2} & \text{when } S_1^2 > S_2^2, \end{cases}$$

which is not Sinha's dominator for $k = 2$ (and $n_1 = n_2$). For the case when the sample sizes are not necessarily equal, Elfessi and Pal give the dominator

$$
\begin{cases}
\hat{\mu}_{GD}(X) & \text{when } S_1^2 \leq S_2^2 \\[2ex]
\dfrac{n_1 \bar{X}_1 + n_2 \bar{X}_2}{n_2 + n_2} & \text{when } S_1^2 > S_2^2,
\end{cases}
$$

which is, when $n_1 = n_2$, Sinha's dominator.

Further results on this problem can be found in Misra and van der Meulen (2005). They consider the case where, for some $k_1 \in \{2, \ldots, k\}$, $0 < \nu_1 \leq \ldots \leq \nu_{k_1}$ and $\nu_i > 0$ for $i = k_1 + 1, \ldots, k$. They show that the Graybill-Deal estimator is, for their restricted parameter space, universally inadmissible and give a dominator. This dominator is obtained by replacing, in the Graybill-Deal estimator, $1/S_i^2$ by V_{k_1-i+1} for $i = 1, \ldots, k_1$, where $V_1 \leq \ldots \leq V_{k_1}$ is a "monotonized version" of $1/S_{k_1}^2, \ldots, 1/S_1^2$. Specifically, the authors took $(V_1, \ldots, V_{k_1})$ to be the minimizer, in $\tau_1, \ldots, \tau_{k_1}$, of

$$
\sum_{i=1}^{k_1} n_{k_1-i+1} \left(\frac{1}{S_{k_1-i+1}^2} - \tau_i \right)^2
$$

under the restriction $\tau_1 \leq \ldots \leq \tau_{k_1}$. Or, to say it another way, $(V_1, \ldots, V_{k_1})$ is the isotonic regression of $(1/S_{k_1}^2, \ldots, 1/S_1^2)$ with weights $n_{k_1}, \ldots, n_1$ and (see, e.g., Barlow, Bartholomew, Bremner and Brunk, 1972, p. 19; or Robertson, Wright and Dykstra, 1988, p. 24; or Chapter 8, Section 8.2) is given by

$$
V_i = \min_{i \leq t \leq k_1} \max_{1 \leq s \leq i} \frac{\displaystyle\sum_{r=s}^{t} \frac{n_{k_1-r+1}}{S_{k_1-r+1}^2}}{\displaystyle\sum_{r=s}^{t} n_{k_1-r+1}}, \qquad i = 1, \ldots, k_1.
$$

For $k = 2$ and $n_1 = n_2$, this Misra–van der Meulen dominator coincides with the Sinha (1979) dominator and thus with the dominator Elfessi and Pal proposed for possibly unequal sample sizes. Further, for the particular case where $k_1 = k$, the Misra–van der Meulen (2005) results can also be found in Misra and van der Meulen (1997).

Elfessi and Pal (1992), Misra and van der Meulen (1997) and Misra and van der Meulen (2005) also show that their universal dominators of the Graybill–Deal estimator dominate it by the Pitman closeness criterion.

Finally, on this Graybill–Deal problem, some earlier results on this problem can be found in Mehta and Gurland (1969). They compare, for $k = 2$ and equal sample sizes, three generalizations of the Graybill-Deal estimator for $\Omega = \{(\mu, \nu_1^2, \nu_2^2) \mid -\infty < \mu < \infty, \nu_1^2 \geq \nu_2^2\}$ as well as for

$\Omega = \{(\mu, v_1^2, v_2^2) \mid -\infty < \mu < \infty, v_1^2 \leq v_2^2\}$. These estimators are of the form $\varphi(F)\bar{X}_1 + (1 - \varphi(F))\bar{X}_2$, where $F = S_2^2/S_1^2$.

Another example where order restrictions on the nuisance parameters make it possible to improve on the estimation of the parameter of interest can be found in Gupta and Singh (1992). They study the case where $X_{i,j} \sim^{ind} \mathcal{N}(\mu_i, v^2)$, $j = 1, \ldots, n_i$, $i = 1, 2$. The paramter of interest is v and the nuissance parameters μ_i satisfy $\mu_1 \leq \mu_2$. The MLE of v is given by

$$\hat{v} = \hat{\sigma}^2 + \frac{n_1 n_2}{(n_1 + n_2)^2}(\bar{X}_1 - \bar{X}_2)^2 I\left(\bar{X}_1 > \bar{X}_2\right),$$

where $\bar{X}_i = \sum_{j=1}^k X_{i,j}/n_i$, $i = 1, 2$ and $\hat{\sigma}^2$ is the unrestricted MLE of v. They show that, for squared-error loss, $\hat{v}$ dominates $\hat{\sigma}^2$.

That using restrictions on the nuissance parameters does not necessarily lead to improved properties of estimators of the parameters of interest is shown by results of Singh, Gupta and Misra (1993). They consider a sample $X_1, \ldots, X_n$ from a population with density $e^{-(x-\mu)/v}$, $x > \mu$ and estimate μ as well as v under the restriction $\mu \leq c$ for a known c and squared-error loss. When estimating v when μ is unknown they find that the unrestricted best (i.e., minimum-risk) affine-equivariant estimator ($\hat{v}_1$, say) and the unrestricted MLE ($\hat{v}_2$, say) are equal. They further show that the restricted MLE ($\hat{v}_3$, say) and $\hat{v}_1$ ($= \hat{v}_2$) have the same risk function. So, by the MSE criterion, these three estimators are equivalent and using the information that $\mu \leq c$ in MLE estimation of v does not improve on the unrestricted MLE ($=$ unrestricted best affine-equivariant).

We now present results on estimating location or scale parameters for k ($k \geq 2$) exponential distributions when all parameters are unknown. Let $X_{i,1}, \ldots, X_{i,n_i}$ be independent with density

$$\frac{1}{v_i}e^{-(x - \mu_i)/v_i} \qquad x > \mu_i, i = 1, \ldots, k.$$

Then the sufficient statistic for (μ_i, v_i) based on (X_i, T_i) is (X_i, T_i), with $X_i = \min_{j=1,\ldots,n_i} X_{i,j}$ and $T_i = \sum_{i=1}^{n_i}(X_{i,j} - X_i)$. The best location-scale equivariant estimator of μ_i based on (X_i, T_i) is $\delta_i^o(X_i) = X_i - (T_i/n_i^2)$, while the one of v_i is T_i/n_i and this last estimator is also the unrestricted MLE of v_i. Further (see Chapter 8, Section 8.2), when $v_1 \leq \ldots \leq v_k$, the MLE of v_i is given by

$$\hat{v}_{MLE,i}(X) = \min_{t \geq i} \max_{s \leq i} \frac{\sum_{r=s}^t T_r}{\sum_{r=s}^t n_r} \qquad i = 1, \ldots, k, \qquad (5.13)$$

and the MLE of μ_i under the restriction $\mu_1 \leq \ldots \leq \mu_k$ is $\hat{\mu}_{MLE,i}(X) = \min(X_i, \ldots, X_k)$. Vijayasree, Misra and Singh (1995), Singh, Gupta and Misra

(1993), Pal and Kushary (1992), as well as Parsian and Sanjari Farsipour (1997) obtain results for the estimation of μ_i and/or ν_i when restrictions are imposed on either the μ_i or the ν_i and all these parameters are unknown. The estimators for which they obtain dominators are (mostly) the best (for their loss function) affine-equivariant ones and the MLEs based on (X_i, T_i) alone. The latter three papers all have $k = 2$, the first three use squared-error loss while the last one uses linex loss.

As an example of this set of results, let the μ_i be simply ordered with the ν_i unknown and unrestricted. Then, when $k = 2$ and the loss is squared-error, $\delta_1^0(X_1) = X_1 - (T_1/n_1^2)$ is an inadmissible estimator of μ_1 and it is dominated by

$$\min(X_1 - T_1/n_1^2, \hat{\mu}_{MLE,1}) \quad \text{by Vijayasree, Misra and Singh (1995)}$$

by

$$
\begin{cases}
X_1 - \dfrac{T_1}{n_1^2} & \text{when } X_1 \le X_2 \\[2ex]
X_2 - \dfrac{1}{n_2^2}\displaystyle\sum_{j=1}^{n_1}(X_{1,j} - X_2) & \text{when } X_1 > X_2 \quad \text{by Singh, Gupta and Misra (1993)}
\end{cases}
$$

and by

$$
\begin{cases}
X_1 - \dfrac{T_1}{n_i^2} & \text{when } X_1 - X_2 \le (T_1/n_1^2) \\[2ex]
X_2 & \text{when } X_1 - X_2 > (T_1/n_1^2) \quad \text{by Pal and Kushary (1992).}
\end{cases}
$$

Pal and Kushary (1992) also look at the case where $\nu_1 = \nu_2 = \nu$ is unknown and $\mu_1 \le \mu_2$. They dominate, for $i = 1, 2$, the estimator

$$\hat{\mu}_{i,c} = X_i - \frac{T_1 + T_2}{n_1(n_1 + n_2 - 1)},$$

which is the minimum-risk-equivariant estimator for estimating μ_i when $\mu_i \le \mu_2$ and $\nu_1 = \nu_2$ based on (X_i, T_1, T_2). For $i = 2$, an example of these dominators is given by

$$
\begin{cases}
\left(X_2 - \dfrac{T_1 + T_2}{n_2(n_1 + n_2 - 1)}\right) & \text{when } X_2 - X_1 \ge \beta(T_1 + T_2) \\[3ex]
\left(X_1 - \dfrac{\gamma(T_1 + T_2)}{n_2(n_1 + n_2 - 1)}\right) & \text{when } X_2 - X_1 < \beta(T_1 + T_2),
\end{cases}
$$

where $\beta = (n_1 - n_2\gamma)/(n_1 n_2(n_1 + n_2 - 1)) > 0$. This dominator does not satisfy the condition that it is less than or equal to X_2 with probabilty 1 for all $\theta \in \Omega$.

Further, by Parsian and Sanjari Farsipour (1997), the best location-equivariant estimator of μ_1 based on (X_1, T_1) alone is given by (see Parsian, Sanjari Farsipour and Nematollahi, 1993)

$$X_1 - \frac{1}{a}\left(\left(\frac{n_1}{n_1 - a}\right)^{(1/n_1)} - 1\right)T_1,$$

when the loss function $L(d, (\mu_1, \nu_1)) = e^{a(d-\mu_1)/\nu_1} - a(d - \mu_1)/\nu_1 - 1$ is used and $a < n_1$, $a \neq 0$. This estimator is, when $k = 2$ and $\mu_1 \leq \mu_2$, dominated by

$$\hat\mu_{MLE,1}(X) - \frac{1}{a}\left(\left(\frac{n_1}{n_1 - a}\right)^{(1/n_1)} - 1\right)\sum_{j=1}^{2}(X_{1,j} - \hat\mu_{MLE,1}(X)).$$

Results for this exponential location-scale case for estimating ν_i when $\nu_1 \leq \ldots \leq \nu_k$ can be found in Vijayasree, Misra and Singh (1995), while both Parsian and Sanjari Farsipour (1997) (for linex loss) and Singh, Gupta and Misra (1993) (for squared-error loss) look at estimating ν_i when $k = 2$ and $\mu_1 \leq \mu_2$. Singh, Gupta and Misra(1993) also use Pitman closeness to compare their estimators and give several examples where a Pitman-closeness comparison of two estimators does not agree with their MSE comparison. For instance, the MLE of ν_1 dominates its unrestricted version by Pitman closeness, but these two estimators are MSE-equivalent. They also have a reversal: for their two MSE-dominators, say, δ_1 and δ_2, of the MLE of ν_1, the MLE dominates δ_1 as well as δ_2 by Pitman closeness. The authors call this non-agreement "paradoxical", but (as noted in Chapter 2) it is known that such reversals occur. Similar results are obtained by Sanjari Farsipour (2002). She looks at the domination results of Parsian and Sanjari Farsipour (1997) and presents pairs of estimators (δ_1, δ_2) for which δ_1 dominates δ_2 by the linex loss function, while by Pitman closeness they are either non-comparable or δ_2 dominates δ_1.

Also, still for this exponential location-scale case, the related problems of estimating (μ_1, μ_2) under the restriction $\mu_1 \leq \mu_2$ and under the restriction $\nu_1/n_1 \leq \nu_2/n_2$ are treated by Jin and Pal (1991). They find dominators for the best location-scale-equivariant estimator (δ_1^o, δ_2^o). They show, e.g., that, for $0 < \alpha \leq 1/2$, the mixed estimator

$$(\delta_{1,\alpha}, \delta_{2,\alpha}) = (\min(\delta_1^o, \alpha\delta_1^o + (1-\alpha)\delta_2^o), \max(\delta_2^o, (1-\alpha)\delta_1^o + \alpha\delta_2^o)) \quad (5.14)$$

dominates (δ_1^o, δ_2^o) when $\mu_1 \leq \mu_2$, while

$$(X_1 - \psi_{1,\alpha}(T_1, T_2), X_2 - \psi_{2,\alpha}(T_1, T_2))$$

dominates (δ_1^o, δ_2^o) when $\nu_1/n_1 \leq \nu_2/n_2$, where

$$\psi_{1,\alpha}(T_1, T_2) = \min\left(\frac{T_1}{n_1^2}, \alpha\frac{T_1}{n_1^2} + (1-\alpha)\frac{T_2}{n_2^2}\right)$$

$$\psi_{2,\alpha}(T_1, T_2) = \max\left(\frac{T_2}{n_2^2}, (1-\alpha)\frac{T_1}{n_1^2} + \alpha\frac{T_2}{n_2^2}\right).$$

Note that the estimator $\max(\delta_2^o, (1-\alpha)\delta_2^o)$ of μ_2 does not satisfy the condition that it is, with probability 1 for all $(\mu_1, \mu_2, \nu_1, \nu_2)$ with $\mu_1 \leq \mu_2$, less than X_2.

Jin and Pal (1991) also have estimators of (μ_1, μ_2) for the case where $\Omega = \{(\mu_1, \mu_2, \nu_1, \nu_2) \mid \mu_1 \leq \mu_2, \nu_1 \leq \nu_2\}$ as well as for the case where $\Omega = \{(\mu_1, \mu_2, \nu_1, \nu_2) \mid \mu_1 \leq \mu_2, \nu_2 \leq \nu_1\}$. Here again, their estimators of μ_2 do not satisfy the condition that they are less than X_2 with probability 1 for all $\mu \in \Omega$.

The results of Jin and Pal (1991) are related to those of Misra and Singh (1994). Each set of authors estimates, for $k = 2$, ordered location parameters of exponential distributions and uses mixed estimators for dominators. These dominators are mixtures of best unrestricted location-scale-equivariant estimators. The difference between the two sets of results is that Misra and Singh have known scale parameters, while Jin and Pal's are unknown. A further difference is that Misra and Singh are interested in the component problem while Jin and Pal are interested in the vector problem. Misra and Singh (1994) do not refer to Jin and Pal (1991).

For Jin and Pal's numerical results, comparing their dominators with (δ_1^o, δ_2^o), see Chapter 7, Section 7.2.

Results on estimating the ratio of the squares of two ordered scale parameters can be found in Kubokawa (1994a). He considers a very general setting of four independent random variables, S_1, S_2, T_1, T_2, where S_i/ν_i, $i = 1, 2$, have a known distribution, whereas the distributions of T_i/ν_i^2, $i = 1, 2$, contain a nuissance parameter. He considers the question of whether an estimator of $\theta = \nu_2^2/\nu_1^2$ based on (S_1, S_2) can, for scale-invariant squared-error loss be improved upon by an estimator based on (S_1, S_2, T_1, T_2). He looks at this question for the unrestricted case, for the case where $\theta \geq 1$ and for the case where $\theta \leq 1$. For the case where $\theta \geq 1$, e.g., he starts out with estimators of the form $\delta_\varphi(S_1, S_2) = \varphi(S_2/S_1)S_2/S_1$ and improves upon them by estimators of the form

$$\delta_{\varphi,\psi}(S_1, S_2, T_2) = \begin{cases} \left(\varphi\left(\dfrac{S_2}{S_1}\right) + \psi\left(\dfrac{T_2}{S_2}\right)\right)\dfrac{S_2}{S_1} & \text{when } T_2 > 0 \\[4mm] \varphi\left(\dfrac{S_2}{S_1}\right)\dfrac{S_2}{S_1} & \text{when } T_2 \leq 0. \end{cases} \tag{5.15}$$

One of the functions φ satisfying his conditions satisfies $\varphi(y) \geq 1/y$, $y > 0$ which guarantees that δ_φ is in $\mathcal{D}$. However, it is not clear to me that there exist (φ, ψ) such that $\delta_{\varphi,\psi}$ is in $\mathcal{D}$ and Kubokawa does not say anything about this question. But in a personal communication to me, he states that, in general, his estimators (5.15) are not in $\mathcal{D}$. Further, and more importantly, it seems to me that the functions ψ he gives below his Theorem 3.2 do not satisfy the conditions of this theorem.

Finally in this section, we look at some results on estimating the smallest variance among $k = 2$ variances based on $Y_{i,j} \sim^{ind} \mathcal{N}(\mu_i, \nu_i)$, $j = 1, \ldots, n_i$, $i = 1, 2$ with the μ_i as well as the ν_i unknown, $\nu_1 \leq \nu_2$ and scale-invariant squared-error loss. Let $\bar{Y}_i = \sum_{j=1}^{n_i} Y_{i,j}/n_i$ and $X_i = \sum_{i=1}^{n_i} (Y_{i,j} - \bar{Y}_i)^2$, $n_i > 1$, $i = 1, 2$. We already saw, in Chapter 5, Section 5.2, that $\delta(X_1) = X_1/(n_1 + 1)$ is the MRE of ν_1 based on X_1 alone and several dominators based on (X_1, X_2) were presented there – mostly in the form of estimating ordered scale parameters of gamma distributions. Ghosh and Sarkar (1994) note that $\delta(X_1)$ can be improved upon by several estimators based on (X_1, W) of the form $(1 - \phi(W))X_1/(n_1 + 1)$, where $W = n_1 \bar{Y}_1^2/X_1$ and they give the following examples of such dominators:

1) Stein (1964, p. 157) has, essentially, a model with $S/\theta \sim \chi_N^2$, $T = \sum_{j=1}^k V_j^2$ where $V_j \sim^{ind} \mathcal{N}(\eta_j, \theta)$, $j = 1, \ldots, k$ and S and T independent. Stein shows, e.g., that for estimating θ in this setting

$$\min\left(\frac{S}{N+2}, \frac{S+T}{N+k+2}\right)$$

dominates $S/(N+2)$. Using this result of Stein in the Ghosh–Sarkar setting with $N = n_1 - 1$, $k = 1$, $S = X_1$ and $T = n_1 \bar{Y}_1^2$, Stein's conditions are satisfied and his result gives that

$$\min\left(\frac{X_1}{n_1 + 1}, \frac{X_1 + n_1 \bar{Y}_1^2}{n_1 + 2}\right)$$

dominates $X_1/(n_1 + 1)$. Note that, with $W = n_1 \bar{Y}_1^2/X_1$, this dominator can be written in the form $(1 - \phi(W))X_1/(n_1 + 1)$ by taking

$$\phi(w) = \max\left(0, \frac{1 - (n_1 + 1)w}{n_1 + 2}\right),$$

the form used by Ghosh and Sarkar (1994, formula (2.4));

2) Strawderman (1974) has the Ghosh and Sarkar (1994) model with $k = 1$ and considers a class of estimators of ν_1^α based on $(X_1, \bar{Y}_1)$. For $\alpha = 1$ these estimators are of the form (see Strawderman, 1974, formula (2.1))

$$\psi\left(\frac{X_1}{X_1 + n_1 \bar{Y}_1^2}\right)(X_1 + n_1 \bar{Y}_1^2)$$

and he shows, e.g., that this estimator with

$$\psi(u) = \frac{u}{n_1 + 1}\left(1 - \varepsilon(u)u^\delta\right)$$

dominates $X_1/(n_1+1)$ as an estimator of ν_1 provided $\varepsilon(u)$ is non-decreasing and $0 \le \varepsilon(u) \le D(\delta)$, where $D(\delta)$ is defined in Strawderman (1974, formula (2.4)). Ghosh and Sarkar (1994) take the special case where $\delta = 1$ and $\varepsilon(u) = \varepsilon > 0$. This gives (see Ghosh and Sarkar, 1994, formula (2.5))

$$\frac{X_1}{n_1+1}\left(1 - \frac{\varepsilon X_1}{X_1 + n_1 \bar{Y}_1^2}\right) \qquad 0 < \varepsilon \le \frac{4(n_1+6)}{(n_1+2)(n_1+3)(n_1+5)}$$

as a dominator of $X_1/(n_1+1)$ for estimating ν_1. Note that this dominator can also be written in the form $(1 - \phi(W))X_1/(n_1+1)$ with

$$\phi(w) = \frac{\varepsilon}{1+w} \qquad 0 < \varepsilon \le \frac{4(n_1+6)}{(n_1+2)(n_1+3)(n_1+5)}; \qquad (5.16)$$

3) Kubokawa (1994b) gives, as a special case of more general results, a class of dominators of $X_1/(n_1+1)$ based on $(X_1, \bar{Y}_1)$ for the Ghosh–Sarkar model with $k = 1$. These estimators are of the form $(1 - \phi(W))X_1/(n_1+1)$ with ϕ any continuously differentiable function satisfying

$$0 < \phi(w) \le 1 - \frac{\mathcal{E}\left(F_1(w\chi_{n_1+1}^2)\right)}{\mathcal{E}\left(F_1(w\chi_{n_1+}^2)\right)},$$

where F_1 is the distribution function of a χ^2 random variable with 1 degree of freedom. This Kubokawa class of dominators contains the generalized Bayes estimators of Brewster and Zidek (1974, Theorem 2.1.4).

The above dominators of $X_1/(n_1+1)$ as an estimator of ν_1 are all based on (X_1, W) only, i.e., on the first sample only. Ghosh and Sarkar (1994) mention several estimators based on (X_1, V), with $V = X_2/X_1$, which dominate $X_1/(n_1+1)$ on $\Omega = \{(\mu_1, \mu_2, \nu_1, \nu_2) \mid \nu_1 \le \nu_2\}$. These dominators are of the form $(1 - \phi(V))X_1/(n_1+1)$ and an example of this class is the class of Strawderman-type dominators of Mathew, Sinha and Sutradhar (1992) with

$$\phi(v) = \frac{\varepsilon}{1+v} \qquad 0 < \varepsilon \le \frac{4(n_2-1)}{(n_1+3)(n_1+n_2)}. \qquad (5.17)$$

Ghosh and Sarkar show that this result can be strengthened to

$$0 < \varepsilon \le \min\left(1, \frac{4(n_2-1)(n_1+n_2+4)}{(n_1+3)(n_1+5)(n_1+n_2)}\right). \qquad (5.18)$$

They also give the following class of dominators of $(1 - \phi(V))X_1/(n_1+1)$ based on (X_1, W, V)

$$\min\left(1 - \phi(V), \frac{(n_1+1)(1+W+V)}{n_1+n_2+1}\right)\frac{X_1}{n_1+1} \qquad (5.19)$$

and the following class of estimators, also based on (X_1, W, V), dominating $(1 - \phi(W))X_1/(n_1+1)$

$$\min\left(1 - \phi(W), \frac{(n_1 + 1)(1 + W + V)}{n_1 + n_2 + 1}\right)\frac{X_1}{n_1 + 1} \tag{5.20}$$

and raise, but do not solve, the question of the existence of an estimator, based on (X_1, W, V), dominating both $(1 - \phi(W))X_1/(n_1 + 1)$ and $(1 - \phi^*(V))X_1/(n_1 + 1)$. Ghosh and Sarkar also mention the Stein-type dominators of Klotz, Milton and Zacks (1969) and Mathew, Sinha and Sutradhar (1992) and they strengthen a result of Mathew, Sinha and Sutradhar for the simple-tree-ordered case with $k > 2$.

Ghosh and Sarkar also give numerical values for the percent decrease in MSE of their estimators relative to the unrestricted MRE of ν_1. Some of these results can be found in Chapter 7, Section 7.2.

5.4 Polygonal, orthant and circular cones

In this section we consider parameter spaces defined by restrictions in the form of more general cones than those defined by inequality restrictions among the parameters.

There are several results on comparing the MLE $c'\hat{\mu}(X)$ of $c'\mu$ with $c'X$ when $X_i \sim^{ind} \mathcal{N}(\mu_i, \nu_i^2)$, $i = 1, \ldots, k$, the ν_i^2 known and Ω is a polygonal cone.

Rueda and Salvador (1995) consider the cone $\Omega = \{\mu \mid a'\mu \geq 0, b'\mu \geq 0\}$, where a and b are known k-dimensional, linearly independent unit vectors. For $k = 2$ they show that, as an estimator of $c'\mu$, $c'\hat{\mu}(X)$ universally dominates $c'X$ for all c. This of course implies that, for all c,

$$\mathcal{E}_\mu(c'(\hat{\mu}(X) - \mu))^2 \leq \mathcal{E}_\mu(c'(X - \mu))^2 \quad \text{for all } \mu \in \Omega, \tag{5.21}$$

i.e., $\hat{\mu}(X)$ is more concentrated about μ than X in the sense of Lehmann (1983, p. 291). For $k > 2$, they do not show universal domination of $c'\hat{\mu}$ over $c'X$, but they do show that (5.21) holds for all c.

For the cone $\Omega = \{\mu \mid a'\mu \geq 0\}$, where a is a known k-dimensional unit vector, Rueda and Salvador (1995) show that, here too, $c'\hat{\mu}(X)$ universally dominates $c'X$ as an estimator of $c'\mu$ for all c and all $\mu \in \Omega$. For this same cone, but now with $a_i \neq 0$ for all $i = 1, \ldots, k$, Rueda, Salvador and Fernández (1997b) compare the distributions of $(|\hat{\mu}_i(X) - \mu_i|, i = 1, \ldots, k)$ and $(|X_i - \mu_i|, i = 1, \ldots, k)$. They show that, for all $t_i > 0$ and all $\mu \in \Omega$,

$$P_\theta(|X_i - \theta_i| \leq t_i, i = 1, \ldots, k) \leq P_\mu(|\hat{\mu}_i(X) - \mu_i| \leq t_i, i = 1, \ldots, k). \tag{5.22}$$

And this implies that, for each $i = 1, \ldots, k$, $\hat{\mu}_i$ universally dominates X_i as an estimator of μ_i, which is Kelly's (1989) result when $k = 2$. For the normal linear model $X = Z\mu + \varepsilon$ with $\varepsilon \sim \mathcal{N}_k(0, I)$, Rueda, Salvador and Fernández (1997a), still for $\Omega = \{\mu \mid a'\mu \geq 0\}$, generalize (5.22) to

$$\left. \begin{array}{c} \text{for all convex } A, \text{ symmetric around zero} \\ P_\mu(|\hat{\mu}(X) - \mu| \in A) \geq P_\mu(|X - \mu| \in A) \quad \text{for all } \mu \in \Omega, \end{array} \right\} \quad (5.23)$$

while, for $Z = I$, Iwasa and Moritani (2002) show that, when $k = 2$, (5.23) holds for Ω convex and closed. Iwasa and Moritani also give generalizations of their result to the case where $k \geq 3$, as well as, for $k \geq 4$, examples where (5.23) does not hold.

A related result for $X_i \sim^{ind} \mathcal{N}(\mu_i, 1)$, $i = 1, \ldots, k$, can be found in Shinozaki and Chang (1999). They have $\Omega = \{\mu \mid \mu_i \geq 0\}$ and show that

$$\mathcal{E}_\mu(c'(\hat{\mu}(X) - \mu))^2 \leq \mathcal{E}_\mu(c'(X - \mu))^2 \quad \text{for all } \mu \in \Omega \quad (5.24)$$

if and only if, for $l = 1, 2$,

$$(\pi + 1) \sum_{i \in S} c_i^2 - \left(\sum_{i \in S} c_i \right)^2 \geq 0 \quad \text{for any } S \subset K_l, \quad (5.25)$$

where $K_1 = \{i \mid c_i > 0\}$ and $K_2 = \{i \mid c_i \leq 0\}$. The condition (5.25) is satisfied for all c if and only if $k \geq 4$.

The Rueda–Salvador, Rueda–Salvador–Fernández and Shinozaki–Chang results have been further generalized by Fernández, Rueda and Salvador (2000). They consider $X = (X_1, \ldots, X_k)$ with a unimodal symmetric density with mean μ and finite variance and the parameter space $\Omega = \{\mu \mid \mu_i \geq 0, i = 1, \ldots, k\}$. They extend the above-mentioned Shinozaki–Chang result to this class of distributions and show that, for (5.24) to hold for all c, it is sufficient for it to hold for $\mu = 0$ and c the central direction of the cone, i.e. for c a vector of ones. For independent samples from $\mathcal{N}(\mu_i, \nu^2)$ distributions, they generalize their results to general orthants and, as one of their examples, give the "increasing-in-average" cone $\Omega = \{\mu \mid \mu_1 \leq (\mu_1 + \mu_2)/2 \leq \ldots \leq (\mu_1 + \ldots + \mu_k)/k\}$ for which $\hat{\mu}(X)$ is more concentrated about μ than X if and only if $k \leq 5$. They give similar results for $X_i \sim^{ind}$ Poisson (μ_i), $i = 1, \ldots, k$ with μ restricted to the cone $\Omega = \{\mu \mid \mu_i \geq a, i = 1, \ldots, k\}$ and $a > 0$ known. Finally, for circular cones and $X_i \sim^{ind} \mathcal{N}(\mu_i, 1)$, Fernández, Rueda and Salvador (1999) compare $c'\hat{\mu}(X)$ with $c'X$. They find that, for any axial angle of the cone, there exists a k such that $c'X$ dominates $c'\hat{\mu}(X)$ at $\mu = 0$ when c is the central direction of the cone. On the other hand, for $k < 4$, e.g., $c'\hat{\mu}$ dominates $c'X$ at $\mu = 0$ for all angles and all c.

None of the above results for Ω a polygonal cone says anything about the universal admissibility of the MLE itself. The only results I have been able to find about this problem is the one by Cohen and Kushary (1998) mentioned in Chapter 3, Section 3.5. They show that, for $X \sim \mathcal{N}(\mu, I)$ with μ restricted to a polygonal cone, the MLE is universally admissible.

5.5 (Admissible) minimax estimators

In this section we report on the (very few) cases where (admissible) minimax estimators are known for restricted estimation problems with nuissance parameters.

Let $X_i \sim^{ind} \mathcal{N}(\mu_i, \nu_i^2)$, $i = 1, 2$, where the ν_i are known and $\Omega = \{\mu \mid \mu_1 \leq \mu_2\}$. The parameter to be estimated is μ_1 and squared-error loss is used. As already seen, the MLE dominates X_1. However, this MLE is inadmissible. This follows from van Eeden and Zidek (2002), who show that $\hat{\mu}_1$ is dominated by

$$\frac{\tau X_1 + X_2}{1 + \tau} - \delta\left(\frac{Z}{1 + \tau}\right),$$

where $Z = X_2 - X_1$, $\tau = \nu_2^2/\nu_1^2$ and δ is a dominator of the MLE of a non-negative normal mean based on a single observation with unit variance. Such dominators can (see Chapter 3, Section 3.4) be found in Shao and Strawderman (1996b).

About minimaxity for this problem: Cohen and Sackrowitz (1970) show that the Pitman estimator of μ_1 (given by $\delta_P(X) = X_1 + \varphi(Z)$ with $Z = X_2 - X_1$ and φ as in the first line of (5.6)) is admissible and minimax and the minimax value is ν_1^2, i.e. the same value as for the unrestricted case. A simpler proof of this minimaxity result can be found in Kumar and Sharma (1988, Theorem 2.3) and a simpler proof of the admissibility is given by van Eeden and Zidek (2002). The fact that δ_P dominates X_1 can also be seen from the following formula for the MSE of δ_P:

$$\mathcal{E}_\mu(\delta_P - \mu_1)^2 = \nu_1^2 - \frac{\nu_1^4}{\sigma^3}(\mu_2 - \mu_1)\, \mathcal{E}_\mu \frac{\phi(Z/\nu)}{\Phi(Z/\nu)},$$

where $\nu^2 = \nu_1^2 + \nu_2^2$. This formula for the MSE of δ_P was proved by Kumar and Sharma (1993) as well as by Al-Saleh (1997) for the case where $\sigma_1^2 = \sigma_2^2 = 1$. The generalization for abitrary variances is given by van Eeden and Zidek (2002).

We also see from the above that X_1 and the MLE are inadmissible minimax, implying that all dominators of them are minimax.

For the normal-mean case with equal known variances and $k = 3$, Kumar and Sharma (1989) show that the first (and thus the third) component of the Pitman estimator of the corresponding components of μ are not minimax when $\Omega = \{\mu \mid \mu_1 \leq \mu_2 \leq \mu_3\}$.

Remark 5.8. Note that Cohen and Sackrowitz's results for the Pitman estimator only hold for the special case they consider, namely the case where $\sigma_2^2 = 1$.

Another case where admissible, as well as minimax estimators, are known is the above normal-mean problem with $\Omega = \{\mu \mid |\mu_2 - \mu_1| \leq c\}$ for a known $c > 0$. Results for this case have been obtained by van Eeden and Zidek (2001, 2004). They show that the Pitman estimator of μ_1, which in this case is given by $\delta_P^*(X) = X_1 + \varphi(Z)$ with $Z = X_2 - X_1$ and φ is in the second line of (5.6), is admissible and dominates X_1. They also show that X_1 and the MLE, given by

$$\hat{\mu}_1(X_1, X_2) = X_1 + \frac{(Z - c)I(Z > c) - (Z + c)I(Z < -c)}{1 + \tau},$$

are inadmissible and that $\hat{\mu}_1$ dominates X_1. A dominator for the MLE is given by

$$\frac{\tau X_1 + X_2}{1 + \tau} - \delta(Z),$$

where $\delta(Z)$ is the projection of

$$\frac{ZI(-c < Z < c) + c(I(Z > c) - I(Z < -c))}{1 + \tau}$$

onto the interval

$$\left[-\frac{c}{1 + \tau} \tanh\left(\frac{c|Z|}{\sigma^2}\right), \frac{c}{1 + \tau} \tanh\left(\frac{c|Z|}{\sigma^2}\right) \right].$$

This dominator is obtained by using results of Moors (1981, 1985).

A minimax estimator of μ_1 when c is small is also obtained by van Eeden and Zidek (2004) . They show that, when $c \leq m_o\sigma$,

$$\delta_{mM}(X_1, X_2) = \frac{\tau X_1 + X_2}{1 + \tau} - \frac{c}{1 + \tau} \tanh\left(\frac{cZ}{\sigma^2}\right),$$

is minimax for estimating μ_1. The minimax value is given by

$$\frac{\sigma_1^2 \nu_2^2}{\nu^4} + \frac{\nu_1^4}{\nu^2} \sup_{|\alpha| \leq m} \mathcal{E}_\alpha (m \tanh(mY) - \alpha)^2,$$

where $m = c/\sigma$, Y is a $\mathcal{N}(\nu, 1)$ random variable and $m_o \approx 1.056742$ is the Casella–Strawderman (1981) constant. Their method of proof is given in Theorem 4.2, where it is used to obtain a minimax estimator of (μ_1, μ_2) when $|\mu_2 - \mu_1| \leq c$.

5.6 Discussion and open problems

In this chapter we looked at questions of admissibility and minimaxity when nuisance parameters are present. Most of the models considered are of the following form: $X_{i,j}$, $j = 1, \ldots, n_i$, $i = 1, \ldots, k$ are independent random variables with distribution function $F_i(x; \mu_i, \nu_i)$ for the $X_{i,j}$, $j = 1, \ldots, n_i$. The μ_i are all unknown, the ν_i are either all known or all unknown and a subvector of the vector (μ, ν) is the parameter of interest, with the rest of the unknown parameters as nuisance parameters.

Most of the questions we looked at are of the form: if, for a given $S \subset \{1, \ldots, k\}$, $\mu_S = \{\mu_i \mid i \in S\}$ (or $(\mu, \nu)_S = \{(\mu_i, \nu_i) \mid i \in S\}$) is the (vector) parameter of interest, can – and if so how – $\{X_{i,j}, j = 1, \ldots, n_i, i \notin S\}$ help improve on a "good" estimator based on $\{X_{i,j}, j = 1, \ldots, n_i, i \in S\}$? Another question is: if δ is a "good" estimator of μ_S when the ν_i are unrestricted, can δ be improved upon on a subset of Ω defined by restrictions on the ν_i? As we have seen, the answers to these questions depend on Ω, Θ, k and the F_i – and, as we have seen, few results have been obtained, even for the relatively simple case of location-scale problems.

A question not touched upon is the relationship between (in)admissibility properties of estimators $\delta_i(X)$ of the components μ_i of μ and these same properties, as an estimator of μ, of the vector $(\delta_1(X), \ldots, \delta_k(X))$. In a more formal setting: let, for $i = 1, \ldots, k$, Θ_i be the projection of Θ onto the μ_i-axis and let the loss function for estimating μ be the sum of the loss functions for the μ_i (i.e. $L(d, \mu) = \sum_{i=1}^{k} L_i(d_i, \mu_i)$). Then, if for each $i = 1, \ldots, k$, $\delta_{o,i}(X)$ and $\delta_i(X)$ are estimators of μ_i for $\mu_i \in \Theta_i$ with $\delta_{o,i}(X)$ dominating $\delta_i(X)$ on Ω, then the vector $\delta(X) = (\delta_1(X), \ldots, \delta_k(X))$ dominates, on Ω, the vector $\delta_o(X) = (\delta_{o,1}(X), \ldots, \delta_{o,k}(X))$ as an estimator of μ provided both δ and δ_o are estimators, i.e., they satisfy

$$P_{\mu,\nu}(\delta(X) \in \Theta) = P_{\mu,\nu}(\delta_o(X) \in \Theta) = 1 \quad \text{for all } (\mu, \nu) \in \Omega.$$

On the other hand, the fact that an estimator δ of $\mu \in \Theta$ is admissible does not imply that its components are admissible as estimators of the corresponding component of μ, but it does imply that at least one of them is admissible.

A related question is: when do estimators $\delta_{o,i}(X)$ which dominate, for each $i = 1, \ldots, k$, $\delta_i(X)$ as an estimator of $\mu_i \in \Theta_i$ give us an estimator $\delta(X) = (\delta_{o,1}(X), \ldots, \delta_{o,k}(X))$ of the vector μ which satisfies (2.3) when $k \geq 2$? A sufficient condition is of course that $\Theta = \prod_{i=1}^{k} \Theta_i$. But in most cases where dominating $\delta_{o,i}$ have been obtained, the resulting δ_o does not satisfy (2.3). And, why should they? Each $\delta_{o,i}$ has been individually constructed to dominate δ_i. Examples of when δ_o is an estimator of μ (and when it is not) can be found in Vijayasree, Misra and Singh (1995). For instance, for the estimation of completely ordered location parameters $\mu_1, \ldots, \mu_k$ of exponential distributions with known scale parameters, Vijayasree, Misra and Singh show that their dominators $\delta_{i,o}$ of the MRE of μ_i (based on X_i alone), satisfy $\delta_{o,1}(X_1) \leq \delta_{o,k}(X_k)$ with probability 1 for all parameter values, while for $(i,j) \neq (1,k)$, $i < j$, $\delta_{o,i}(X_i) \leq \delta_{o,j}(X_j)$ holds with probability < 1 for some parameter values. This result of course implies that, when $k = 2$, their δ_o satisfies (2.3). On the other hand, the Vijayasree–Misra–Singh dominator of the restricted MLE of μ_i does satisfy (2.3) for all k. Of course, if one is really only interested in estimating a component of μ, one would want the class of estimators to choose from to be as large as possible, i.e., one would not want it to be restricted by the requirement that it, together with dominators of the other components (those one is not interested in), leads to an estimator of the vector.

6

The linear model

There are a large number of papers which are concerned with estimation problems in restricted parameter spaces for the linear model

$$X = Z\theta + \varepsilon, \tag{6.1}$$

where X is an $n \times 1$ vector of observables, Z is a given nonstochastic $n \times k$ matrix of rank k, θ is a $k \times 1$ vector of unknown parameters and ε is a $n \times 1$ vector of random variables with $\mathcal{E}\varepsilon = 0$ and $cov(\varepsilon) = \Sigma$, known or unknown. In the latter case Σ is either a matrix of nuisance parameters, or it is estimated and this estimator is then used in the estimation of θ. The problem considered in such papers is the estimation of θ, or a subvector of θ with the other components as nuisance parameters, when θ is restricted to a closed convex subset of R^k. The estimators are often linear estimators, i.e. they are of the form $\hat{\theta}(X) = AX + b$. Such estimators generally do not satisfy (2.3), i.e. the authors consider (see Chapter 2) the $(\mathcal{D}_o, \Theta)$-problem and not the $(\mathcal{D}, \Theta)$-problem. Papers on this subject which are (almost) exclusively concerned with $(\mathcal{D}_o, \Theta)$-problems are not discussed, but the ones I have found are listed in the references and indicated with an asterisk before the first author's name.

Results for $(\mathcal{D}, \Theta)$-problems for the model (6.1) with $\Theta = \{\theta \mid b'\theta \geq r\}$ for a given $k \times 1$ vector b and a given constant r can be found in Section 6.1. In Section 6.2 two kinds of restrictions on θ are considered, namely, the case where $r_1 \leq b'\theta \leq r_2$, as well as the case where $k = 2$ and $\theta_i \geq 0$, $i = 1, 2$. Section 6.3 presents results of Moors and van Houwelingen (1993) for simultaneous interval restrictions on the components of $A\theta$, where A is a known $k \times k$ positive definite matrix, as well as for ellipsoidal restrictions. Most of the results in this chapter hold for squared-error loss and MSE will stand for mean-squared-error.

We note that several of the results presented in Section 6.1 are related to, and sometimes overlap with, those of Chang (1981, 1982), Sengupta and Sen (1991) and Kuriki and Takemura (2000) – results which are presented in Chapter 3, Section 3.4.

6.1 One linear inequality restriction

For the model (6.1) and $\Theta = \{\theta \mid b'\theta \geq r\}$, Lovell and Prescott (1970) as well as Thomson and Schmidt (1982) consider the special case where $b' = (1, 0, \ldots, 0)$, $\Sigma = \sigma^2 I$ and (without loss of generality) $r = 0$. Their estimators of θ are the least-squares estimators (LSE), i.e., their estimators $\hat{\theta}(X)$ minimize, for $\theta \in \Theta$, $\sum_{i=1}^{n}(X_i - \sum_{j=1}^{k} z_{i,j}\theta_j)^2$. If $\delta^*(X)$ denotes the unrestricted least-squares estimator (URLSE) of θ then $\hat{\theta}(X) = \delta^*(X)$ when $\delta_1^*(X) \geq 0$ and $\hat{\theta}(X)$ is the URLSE for the model (6.1) with θ replaced by $(0, \theta_2, \ldots, \theta_k)$ when $\delta_1^*(X) < 0$. The authors compare, for $i = 1, \ldots, k$, $\hat{\theta}_i$ with δ_i^* for squared-error loss. Obviously, $\hat{\theta}_1$ dominates δ_1^* unless $P_\theta(\delta_1^*(X) < 0) = 0$ for all θ with $\theta_1 \geq 0$. However, as Lovell and Prescott show, $\hat{\theta}_i$ does not necessarily dominate δ_i^* for $i \geq 2$. They give, for $k = 2$, an example of a matrix Z and a distribution for ε for which δ_2^* dominates $\hat{\theta}_2$. But they also show, using explicit formulas for the MSE's, that $\hat{\theta}_i$ dominates δ_i^* on $\{\theta \mid \theta_1 \geq 0\}$ for $i = 1, \ldots, k$ when $\varepsilon \sim \mathcal{N}_n(0, \sigma^2 I)$. Thomson and Schmidt generalize this last result of Lovell and Prescott. They note that Lovell and Prescott's formula (4.4) for the MSE of $\hat{\theta}_i$ when $\varepsilon \sim \mathcal{N}_n(0, \sigma^2 I)$, can be written in the form

$$\mathrm{MSE}(\hat{\theta}_i) = \sigma_i^2 + \rho_{i,1}^2 \sigma_i^2((c^2 - 1)\Phi(c) + c\phi(c)) \quad i = 1, \ldots, k, \qquad (6.2)$$

where σ_i^2 is the variance of δ_i^*, $\rho_{i,1}$ is the correlation coefficient between δ_i^* and δ_1^*, Φ and ϕ are, respectively, the standard normal distribution function and density and $c = -\theta_1/\sigma_1$. This then shows that, for $\varepsilon \sim \mathcal{N}(0, \sigma^2 I)$, $\hat{\theta}_i$ dominates δ_i^* on the larger space $\{\theta \mid \theta_1 \geq -.84\sigma_1\}$ because $\Phi(c)(c^2 - 1) + c\phi(c)$ is positive for $c > .84$ and negative for $c < .84$. It also shows that the MSE's depend on θ only through θ_1.

More results for the case where $\Theta = \{\theta \mid b'\theta \geq r\}$ can be found in Judge, Yancey, Bock and Bohrer (1984), Wan (1994b), Ohtani and Wan (1998) and Wan and Ohtani (2000). Each of these four sets of authors assume $\varepsilon \sim \mathcal{N}_n(0, \sigma^2 I)$ and follow Judge and Yancey (1981) in reparametrizing the problem as follows. Let $V = Z(Z'Z)^{-1/2}Q$ where Q is an orthogonal matrix such that $b'(Z'Z)^{-1/2}Q = (1, 0, \ldots, 0)$ and let $\beta = Q'(Z'Z)^{1/2}\theta$. Then $X = V\beta + \varepsilon$ and the parameter space becomes $\{\beta \mid \beta_1 \geq r\}$, which they take, without loss of generality, to be $\{\beta \mid \beta_1 \geq 0\}$. Further, the equality-restricted least-squares estimator of θ, i.e., the least-squares estimator of θ under the restriction $b'\theta = 0$, is given by (see Judge and Yancey, 1981)

$$\hat{\theta}_{ER}(X) = \delta^*(X) - (Z'Z)^{-1}b(b'(Z'Z)^{-1}b)^{-1}b'\delta^*(X),$$

where $\delta^*(X)$ is the unrestricted least-squares estimator of θ. Now, using the fact that $b'(Z'Z)^{-1}b = b'(Z'Z)^{-1/2}QQ'(Z'Z)^{-1/2}b = 1$ gives

$$\hat{\theta}_{ER}(X) = \delta^*(X) - (Z'Z)^{-1}bb'\delta^*(X).$$

Further, the unrestricted least-squares estimator $\delta(X)$ of β is given by

$$\delta(X) = (V'V)^{-1}V'X = V'X,$$

so the equality-restricted least-squares estimator of $\beta = Q'(Z'Z)^{1/2}\theta$ is given by

$$\hat{\beta}_{ER}(X) = \delta(X) - Q'(Z'Z)^{-1/2}bb'\delta^*(X)$$

$$= \delta(X) - Q'(Z'Z)^{-1/2}bb'(Z'Z)^{-1/2}Q\delta(X)$$

$$= (0, \delta_2(X), \ldots, \delta_k(X))',$$

because

$$\delta^*(X) = (Z'Z)^{-1/2}Q\delta(X) \text{ and } b'(Z'Z)^{-1/2}Q = (1, 0, \ldots, 0).$$

The inequality-restricted least-squares estimator, $\hat{\beta}(X)$, of β then becomes

$$\hat{\beta}(X) = \begin{cases} \begin{pmatrix} 0 \\ \delta_{(k-1)}(X) \end{pmatrix} & \text{when } \delta_1(X) < 0 \\[2ex] \delta(X) & \text{when } \delta_1(X) \geq 0, \end{cases} \tag{6.3}$$

where $\delta_{(k-1)}(X) = (\delta_2(X), \ldots, \delta_k(X))'$. Of course, when $\varepsilon \sim \mathcal{N}_n(0, \sigma^2 I)$, the above least-squares estimator are the corresponding MLE's.

Judge, Yancey, Bock and Bohrer (1984), assuming $\varepsilon \sim \mathcal{N}_n(0, \sigma^2 I)$ and σ^2 known, estimate β and, for squared-error loss, present several estimators dominating its MLE $\hat{\beta}$ under the restriction $\beta_1 \geq 0$. An example of their dominators is the Stein inequality-restricted estimator $\hat{\beta}_{SIR}$ obtained from $\hat{\beta}$ (see (6.3)) by replacing $\delta_{(k-1)}$ and δ by appropriate James–Stein dominators. This gives the following expression for $\hat{\beta}_{SIR}(X)$:

$$\begin{cases} \begin{pmatrix} 0 \\ \left(1 - \dfrac{c_1\sigma^2}{\delta'_{(k-1)}(X)\delta_{(k-1)}(X)}\right)\delta_{(k-1)}(X) \end{pmatrix} & \text{when } \delta_1(X) < 0 \\[3ex] \left(1 - \dfrac{c_2\sigma^2}{\delta'(X)\delta(X)}\right)\delta(X) & \text{when } \delta_1(X) \geq 0, \end{cases} \tag{6.4}$$

where $k \geq 4$, $0 < c_1 < 2(k-3)$ and $0 < c_2 < 2(k-2)$. The choice of (c_1, c_2) minimizing the risk function of $\hat{\beta}_{SIR}$ is $c_1 = k-3$, $c_2 = k-2$. A second example of their dominators is a positive Stein inequality-restricted estimator given by

$$\hat{\beta}_{PSIR}(X) = \begin{cases} \begin{pmatrix} 0 \\ \hat{\beta}_{PS,(k-1)}(X) \end{pmatrix} & \text{when } \delta_1(X) < 0 \\ \\ \hat{\beta}_{PS}(X) & \text{when } \delta_1(X) \geq 0, \end{cases} \tag{6.5}$$

where $\hat{\beta}_{PS}(X) = (1 - c\sigma^2/(\delta'(X)\delta(X)))I(\delta'(X)\delta(X) > c\sigma^2)\delta(X)$ is the (unrestricted) positive Stein estimator, $0 < c < 2(k-2)$ and $\hat{\beta}_{PS,(k-1)}(X) = (\hat{\beta}_{PS,2}(X), \ldots, \hat{\beta}_{PS,k}(X))$. This $\hat{\beta}_{PSIR}$ also dominates $\hat{\beta}_{SIR}$. As a third dominator of $\hat{\beta}$ they have a mixed MLE-Stein inequality-restricted estimator given by

$$\hat{\beta}_{MLES}(X) = \begin{cases} \begin{pmatrix} 0 \\ \delta_{(k-1)}(X) \end{pmatrix} & \text{when } \delta_1(X) < 0 \\ \\ \left(1 - \dfrac{c\sigma^2}{\delta'(X)\delta(X)}\right)\delta(X) & \text{when } \delta_1(X) \geq 0, \end{cases} \tag{6.6}$$

with $0 < c < 2(k-2)$. The authors give MSE-formulas for the vectors $\hat{\beta}$ and $\hat{\beta}_{SIR}$, but not for their components. However, formulas for the other vectors and all components can easily be obtained by the authors' methods. For $\hat{\beta}$, e.g., they show that

$$\text{MSE}(\hat{\beta}) = \text{MSE}(\delta) + \sigma^2 \mathcal{E}_\beta \left(\frac{\beta_1^2}{\sigma^2} - \left(\frac{\delta_1(X) - \beta_1}{\sigma}\right)^2\right) I(\delta_1(X) < 0),$$

with $\text{MSE}(\delta) = k\sigma^2$ because $\delta_i(X) \sim^{ind} \mathcal{N}(\beta_i, \sigma^2)$, $i = 1, \ldots, k$. From this formula and the definition (6.3) of $\hat{\beta}$, it follows that

$$\frac{\text{MSE}(\hat{\beta}_1)}{\sigma^2} = 1 + \mathcal{E}\left(\frac{\beta_1^2}{\sigma^2} - \left(\frac{\delta_1(X) - \beta_1}{\sigma}\right)^2\right) I(\delta_1(X) < 0),$$

and

$$\frac{\text{MSE}_\beta(\hat{\beta}_i)}{\sigma^2} = 1 \quad i = 2, \ldots, k,$$

with $\mathcal{E}_\beta \left(\dfrac{\beta_1^2}{\sigma^2} - \left(\dfrac{\delta_1(X) - \beta_1}{\sigma}\right)^2\right) I(\delta_1(X) < 0) = (c^2 - 1)\Phi(c) + c\phi(c)$ and $c = -\beta_1/\sigma$. This of course gives the Lovell–Prescott–Thomson–Schmidt formula (6.2) for $\text{MSE}(\hat{\beta}_1)$ for the case where $Z'Z = I$.

For the difference between the risk functions of $\hat{\beta}_{SIR}$ and $\hat{\beta}$ they find

$$\text{MSE}(\hat{\beta}_{SIR}, \beta) - \text{MSE}(\hat{\beta}, \beta) =$$

$$c_1(c_1 - 2(k-3))P_\beta(\delta_1(X) < 0)\mathcal{E}\frac{1}{\chi^2_{k-1,\lambda}} +$$

$$c_2(c_2 - 2(k-2))\mathcal{E}_\beta I(\delta_1(X) > 0)\frac{1}{\delta'(X)\delta(X)},$$

where $\chi^2_{k-1,\lambda}$ is a χ^2 random variable with $k-1$ degrees of freedom and non-centrality parameter $\lambda = \beta'_{(k-1)}\beta_{(k-1)}/2$. So, $\hat{\beta}_{SIR}$ dominates $\hat{\beta}$ when $0 < c_1 < 2(k-3)$ and $0 < c_2 < 2(k-2)$. The authors also show that $\hat{\beta}_{MLES}$ dominates $\hat{\beta}$.

The model of Judge, Yancey, Bock and Bohrer (1984) is a special case of the one of Chang (1981), but there does not seem to be any overlap between their classes of dominators. There is, however, an overlap of their results with those of Sengupta and Sen (1991).

Wan (1994b), assuming $\varepsilon \sim \mathcal{N}_n(0, \sigma^2 I)$ and $\beta_1 \geq 0$, estimates β by its MLE $\hat{\beta}$ as well as by an inequality-restricted pretest estimator $\hat{\beta}_{IPT}$, say. This pretest uses $(X - V\delta(X))'(X - V\delta(X))/(n-k)$ as an estimator of σ^2. To compare these estimators he uses the so-called balanced loss function given by

$$L(d, \beta) = w(X - Vd)'(X - Vd) + (1 - w)(d - \beta)'(d - \beta),$$

where $w \in [0, 1]$ is a given number. The first term in $L(d, \beta)$ is a measure of goodness-of-fit of the model, while the second term is a weighted squared-error loss for an estimator of β.

Wan gives explicit expressions for the expected losses of his estimators and graphically illustrates comparisons between $\hat{\beta}$, $\hat{\beta}_{IPT}$ and δ as a function of β_1/σ for $n = 30$, $k = 5$ and $w = 0.0, 0.3$ and 0.5. In each graph with $w < 1/2$, $\hat{\beta}$ dominates $\hat{\beta}_{IPT}$ and $\hat{\beta}_{IPT}$ dominates δ. From these graphs it can also be seen (analogous to the results of Lovell–Prescott–Thomson–Schmidt model) that the domination results hold on a space larger than $\{\beta \mid \beta_1 \geq 0\}$. For $w = 1/2$, $\hat{\beta}$ and $\hat{\beta}_{IPT}$ are about equivalent and dominate δ, while for $w > 1/2$, δ dominates $\hat{\beta}_{IPT}$ which dominates $\hat{\beta}$ on a space larger than $\{\beta \mid \beta_1 \geq 0\}$, but on a large part of $\{\beta \mid \beta_1 \geq 0\}$ there is no difference between the three. Judge, Yancey, Bock and Bohrer (1984) have a graph of the risks of their estimators as a function of β_1/σ for $k = 4$ and $\beta = (1, 0, 0, 0)'$. Here too, the domination results hold on a space larger than $\{\beta \mid \beta_1 \geq 0\}$.

Ohtani and Wan (1998), assuming $\varepsilon \sim \mathcal{N}_n(0, \sigma^2 I)$, estimate β when σ^2 is unknown and $\beta_1 \geq 0$. They start with the estimator (6.4) and, for $\delta_1(X) \geq 0$, replace σ^2 by its Stein (1964) variance estimator given by

$$\tilde{\sigma}_S^2 = \min\left(\frac{e'(X)e(X)}{n-k+2}, \frac{X'X}{n+2}\right),$$

where $e(X) = X - V\delta(X)$. For $\delta_1(X) < 0$ they replace σ^2 by its equality-restricted (i.e., assuming $\beta_1 = 0$) Stein variance estimator given by

$$\hat{\sigma}_S^2 = \begin{cases} \dfrac{e'(X)e(X) + \delta_1^2(X)}{n-k+3} & \text{when } \dfrac{\delta'(X)\delta(X)}{k-1} \geq \dfrac{\tilde{e}'(X)\tilde{e}(X)}{n-k+3} \\[3mm] \dfrac{X'X}{n+2} & \text{when } \dfrac{\delta'(X)\delta(X)}{k-1} < \dfrac{\tilde{e}'(X)\tilde{e}(X)}{n-k+3}, \end{cases}$$

where $\tilde{e}(X) = X - V(0, \delta'_{(k-1)})'$. They justify these estimators through a pretest argument and give results of numerical comparisons, using scaled squared-error loss $(=\text{MSE}/\sigma^2)$, of their estimator with the estimator $\hat{\beta}_{SIR}$ for $n = 20$, $k = 5$, 10, 15, $\beta'_{(k-1)}\beta_{(k-1)}/\sigma^2 = .1$, 1.0, 6.0 and $\beta_1/\sigma = -3.0, -2.5, \ldots, 3.0$. For each combination of these values of $\beta'_{(k-1)}\beta_{(k-1)}/\sigma^2$, β_1/σ and k, they find that their estimator has a smaller risk than does $\hat{\beta}_{SIR}$. They do not have an analytical comparison of these estimators, but show analytically that both risks converge to ∞ as $\beta_1 \to -\infty$ and both risks converge to k, the constant risk of δ, when $\beta_1 \to \infty$. For the unrestricted case Berry (1994) shows that, for estimating a multivariate normal mean, the James–Stein estimator can be improved upon by incorporating the Stein (1964) variance estimator.

Wan and Ohtani (2000) study an adaptive estimator of β when $\beta_1 \geq 0$. They start out with the class of (unrestricted) adaptive estimators given by

$$\delta_F(X) = \frac{\delta'(X)\delta(X)}{d_1 e'(X)e(X) + \delta'(X)\delta(X)}\delta(X).$$

Farebrother (1975) and Ohtani (1996) each study special cases of these estimators with, respectively, $d_1 = 1/(n-k)$ and $d_1 = k/(n-k)$. Wan and Ohtani then propose an inequality-restricted version of $\delta_F(X)$ given by

$$\begin{cases} \delta_F(X) & \text{when } \delta_1(X) \geq 0 \\[3mm] \left(0, \dfrac{\delta'(X)\delta(X)}{d_2(e'(X)e(X) + \delta_1^2(X)) + \delta'_{(k-1)}\delta_{(k-1)}(X)}\delta'_{(k-1)}\right)' & \text{when } \delta_1(X) < 0. \end{cases}$$

Using scaled squared-error loss, the authors give an explicit expression for the risk function of their estimator and give analytical comparisons as well as numerical ones with the MLE $\hat{\beta}$ and the Judge, Yancey, Bock and Bohrer estimator (6.4). Analytically they find that $\text{MSE}(\hat{\beta}_F, \beta) \leq \text{MSE}(\hat{\beta}, \beta)$ for all

$\beta_1 \leq 0$ when $d_1 \geq 0$, $d_2 \geq 0$ and $\beta'_{(k-1)}\beta_{(k-1)} \leq (k-1)\sigma^2/2$. So their estimator does, relative to the MLE, better outside the parameter space than inside it. The authors also have numerical comparisons for various combinations of values of n, k, β_1/σ and $\beta'_{(k-1)}\beta_{(k-1)}/\sigma^2$ between the MLE $\hat{\beta}$, $\hat{\beta}_{SIR}$ and two special cases of their estimator: namely, $\hat{\beta}_{F,1}$ with $d_1 = 1/(n-k)$ and $d_2 = 1/(n-k+1)$ and $\hat{\beta}_{F,2}$ with $d_1 = k/(n-k)$ and $d_2 = (k-1)/(n-k+1)$. These results indicate that, for $\beta_1 \geq 0$, (i) when $k = 2$ (where $\hat{\beta}_{SIR}$ is not available) with $n = 15$ as well as with $n = 40$, there is not much difference between $\hat{\beta}_{F,1}$ and $\hat{\beta}_{F,2}$ and the best of the three estimators is $\hat{\beta}_{F,2}$; (ii) when $k = 8$ with $n = 15$ as well as with $n = 40$, $\hat{\beta}_{F,2}$ is clearly preferable over $\hat{\beta}_{F,1}$ as well as over $\hat{\beta}_{SIR}$; (iii) When $k = 25$ with $n = 40$ as well as with $n = 100$, $\hat{\beta}_{F,2}$ is again preferrred over $\hat{\beta}_{F,1}$, but the best is $\hat{\beta}_{SIR}$.

6.2 Interval and quadrant restrictions

Escobar and Skarpness (1987) and Wan (1994a) start with the model (6.1) with $\varepsilon \sim \mathcal{N}(0, \sigma^2 I)$ and σ^2 known. They restrict θ to $r_1 \leq b'\theta \leq r_2$ for a given $k \times 1$ vector b and given numbers $r_1 < r_2$ and transform the model as follows. Let D be a $k \times (k-1)$ matrix such that $B' = (b, D)$ is non-singular and let $V = ZB^{-1}$. Then the model $X = Z\theta + \varepsilon$ becomes $X = V\beta + \varepsilon$ with $\beta = B\theta$ and the restricted parameter space becomes $\{\beta \mid r_1 \leq \beta_1 \leq r_2\}$. Let, as before, $\delta(X) = (V'V)^{-1}V'X$ be the unrestricted MLE of β. Then (see Klemm and Sposito, 1980) the restricted MLE $\hat{\beta}$ of β is given by

$$\hat{\beta}(X) = \delta(X) + \gamma(\hat{\beta}_1(X) - \delta_1(X))$$

where $\gamma = (V'V)^{-1}u/(u'(V'V)^{-1}u)$, $u' = (1, 0, \ldots, 0)'$ and

$$\hat{\beta}_1(X) = \begin{cases} r_1 & \text{when } \delta_1(X) < r_1 \\ \delta_1(X) & \text{when } r_1 \leq \delta_1(X) \leq r_2, \\ r_2 & \text{when } \delta_i(X) > r_2. \end{cases}$$

Now, write $V = (V_1, V_2)$ and $\gamma' = (\gamma_1, \Gamma'_2)$ where V_1 is an $n \times 1$ vector, V_2 is an $n \times (k-1)$ matrix, $\gamma_1 = 1$ is the first element of the $k \times 1$ vector γ and $\Gamma_2 = (\gamma_2, \ldots, \gamma_k)' = -(V'_2 V_2)^{-1}V'_2 V_1$. Then, with $\hat{\beta}(X) = (\hat{\beta}_1(X), \hat{\beta}'_{(k-1)}(X))'$, Escobar and Skarpness show that $\hat{\beta}_{(k-1)}(X)$ can also be written in the form

$$\hat{\beta}_{(k-1)}(X) = \delta_{(k-1)}(X) + \gamma_2(\hat{\beta}_1(X) - \delta_1(X)). \tag{6.7}$$

Remark 6.1. The transformation of the model (6.1) used by Escobar and Skarpness (1987) is not necessarily the Judge and Yancey (1981) one. So, *the Escobar–Skarpness parameter β is not necessarily the one obtained by using the Judge–Yancey transformation. They are the same parameter (and thus have, under the same restrictions on θ, the same least-squares estimator) when $V'V = I$. And in that case $\gamma = (1, 0, \ldots, 0)'$ and this implies that $\hat{\beta}_{(k-1)}(X) = \delta_{(k-1)}(X)$.*

Escobar and Skarpness (1987) compare $\hat{\beta}$ with δ using squared-error loss. Obviously, as for the Lovell–Prescott (1970) model, $\hat{\beta}_1$ dominates δ_1 on $\{\beta \mid r_1 \le \beta_1 \le r_2\}$ unless $P_\beta(r_1 \le \delta_1(X) \le r_2) = 1$ for all β with $\beta_1 \in [r_1, r_2]$. The authors' robustness results with respect to misspecification of Θ are presented in Chapter 7, Section 7.2.

For the comparison of $\hat{\beta}_i$ with δ_i for $i = 2, \ldots, k$, Escobar and Skarpness show that
$$\mathrm{MSE}(\delta_i) - \mathrm{MSE}(\hat{\beta}_i) = \gamma_i^2(\mathrm{MSE}(\delta_1) - \mathrm{MSE}(\hat{\beta}_1))$$
and note that in the proof of this result the normality assumption of the residuals plays a central role. So, what these authors show is that, for $\varepsilon \sim \mathcal{N}_n(0, \sigma^2 I)$ and all β with $r_1 \le \beta_1 \le r_2$, $\mathrm{MSE}(\hat{\beta}_i) < \mathrm{MSE}(\delta_i)$ for any $i = 2, \ldots, k$ for which $\gamma_i \ne 0$. And, from (6.7), that $\hat{\beta}_i(X) = \delta_i(X)$ with probability 1 for all β with $r_1 \le \beta_1 \le r_2$ for any $i \ge 2$ with $\gamma_i = 0$. Escobar and Skarpness do not have an example, like Lovell and Prescott (1970) do, of a distribution for ε for which δ_i dominates $\hat{\beta}_i$ for some $i \ge 2$.

The Escobar–Skarpness model and transformation, but then with a multivariate t distribution for ε, is considered by Ohtani (1991). He gives explicit expressions, as well as graphs, for both the bias and the MSE of the least-squares estimator of β_1.

More results for the model (6.1) with $\varepsilon \sim \mathcal{N}_n(0, \sigma^2 I)$, known σ^2 and $r_1 \le b'\theta \le r_2$ can be found in Wan (1994a). He uses the Judge and Yancey (1981) transformation and compares, for squared-error loss, several estimators of β_1: namely, the unrestricted MLE, the MLE, a pretest estimator of Hasegawa (1991) (who compares it with the MLE) and several "Stein-adjusted" version of them which are analogues of the Judge–Yancey–Bock–Bohrer estimators (6.4)–(6.6). For some of these estimators he has analytical results, for all of them he gives graphs of their risks as a function of β_1 for several combinations of values of $k \ge 4$, $(r_2 - r_1)/\sigma$, the α of the pretest and $\beta'_{(k-1)}\beta_{(k-1)}/\sigma^2$. For several of the comparisons the domination result holds on a space larger than $\{\beta \mid \beta_1 \ge 0\}$.

Thomson (1982) considers the model (6.1), where $Z = (Z_1, Z_2)$ with, for $i = 1, 2$, Z_i an $n \times 1$ vector. The 2×1 vector θ is restricted to $\theta_i \ge 0$, $i = 1, 2$ and he supposes $\varepsilon \sim \mathcal{N}_n(0, \sigma^2 I)$. The estimators of θ he considers are (in the notation used above for the Lovell–Prescott–Thomson–Schmidt model) the unrestricted MLE δ^*, the MLE $\hat{\theta}$ and the equality-restricted MLE $\hat{\theta}_{ER}$ under the restrictions $\theta_1 = \theta_2 = 0$. Thomson studies the bias and the MSE of these three estimators analytically as well as graphically.

For the MSE of $\hat{\theta}_1$ he finds

$$
\left.
\begin{aligned}
&\frac{\text{MSE}(\hat{\theta}_1)}{\sigma_1^2} \\[2mm]
&= c_1^2 + (1 - c_1^2)F_\rho(-c_1, -c_2) + c_1\phi(c_1)\Phi(d_2) \\[2mm]
&\quad + \rho^2 c_2 \phi(c_2)\Phi(d_1) + \rho\sqrt{\frac{1-\rho^2}{2\pi}}\phi\left(\sqrt{\frac{c_1^2 - 2\rho c_1 c_2 + c_2^2}{1-\rho^2}}\right) \\[2mm]
&\quad + (1 - \rho^2 - c_1^2 + \rho^2 c_2^2)\Phi(c_2)\Phi(d_1) + \sqrt{1 - \rho^2}(c_1 + \rho c_2)\phi(d_1)\Phi(c_2),
\end{aligned}
\right\} \quad (6.8)
$$

where $d_1 = (\rho c_2 - c_1)/\sqrt{1 - \rho^2}$, $d_2 = (\rho c_1 - c_2)/\sqrt{1 - \rho^2}$, $c_i = -\theta_i/\sigma_i$, $i = 1, 2$, σ_i^2 is the variance of δ_i^*, $i = 1, 2$, ρ is the correlation coefficient of δ_1^* and δ_2^* and F_ρ is the distribution function of the standard two-dimensional normal distribution with correlation coefficient ρ.

For the special case where $\rho = 0$, i.e., where $Z_1' Z_2 = 0$, this gives

$$
\frac{\text{MSE}(\hat{\theta}_1)}{\sigma_1^2} = 1 + (c_1^2 - 1)\Phi(c_1) + c_1\phi(c_1),
$$

which is the Lovell–Prescott–Thomson–Schmidt formula (6.2) with $i = 1$ for the case of exactly one constraint (namely, $\theta_1 \geq 0$). So, as Thomson notes, when $\rho = 0$, the MLE of θ_1 under the double constraint ($\theta_1 \geq 0$, $\theta_2 \geq 0$) has the same MSE as the MLE of θ_1 under the single constraint $\theta_1 \geq 0$. And using the Thomson–Schmidt result, we then see that under the single constraint the MLE $\hat{\theta}_1$ of θ_1 dominates δ_1^* for all (θ_1, θ_2) with $c_1 \geq -.84$ when when $\rho = 0$. Further, using Thomson's formula for the bias of $\hat{\theta}_1$, we see that, when $\rho = 0$,

$$
\mathcal{E}_\theta \hat{\theta}_1 - \theta_1 = (c_1 + \phi(c_1) - c_1\Phi(-c_1))\sigma_1
$$

which is the formula for the bias of $\hat{\theta}_1$ in the Lovell–Prescott–Thomson–Schmidt single-constraint model, because in that model $\hat{\theta}_1 = \max(0, \delta_1^*(X))$. So, when $\rho = 0$, the MLE's of θ_1 in these two models have the same distribution.

When $\rho \neq 0$ the problem of comparing the estimators is much more complicated. In fact, as can be seen from (6.8), the MSE of $\hat{\theta}_1/\sigma_1$ depends on θ_1/σ_1, θ_2/σ_2 and ρ. The author makes graphical comparisons between the MSEs of his estimators. In each one of these graphs, $\hat{\theta}_1$ dominates δ_1^* on a set larger than $\Theta = \{\theta \mid \theta_1 \geq 0, \theta_2 \geq 0\}$, but he does not say anything about whether this dominance holds in general. For the more general case where $k > 2$ and $\theta_i \geq 0$ for $i = 1, \ldots, k$ the problem gets of course even more difficult.

6.3 Results of Moors and van Houwelingen for polyhedral and ellipsoidal restrictions

As already mentioned in Chapter 3, Section 3.2, Moors and van Houwelingen (1993) apply Moors' (1981, 1985) results to the linear model (6.1) with $\varepsilon \sim \mathcal{N}_n(0, \sigma^2 I)$.

They consider two parameter spaces:

1) $\Theta = \{\theta \mid a \leq A\theta \leq b\}$, where $a = (a_1, \ldots, a_k)$ and $b = (b_1, \ldots, b_k)$ are known $k \times 1$ vectors, A is a known $k \times k$ positive definite matrix and the inequalities are component-wise inequalities. When $-\infty < a_i < b_i < \infty$, the authors suppose, without loss of generality, that $a_i = -b_i$ and, when $a_i = -\infty < b_i < \infty$ (resp. $-\infty < a_i < b_i = \infty$), they suppose, without loss of generality, that $b_i = 0$ (resp. $a_i = 0$). Putting $\beta = A\theta$ and $V = ZA^{-1}$, their model becomes $X = V\beta + \varepsilon$ and their parameter space becomes $C = \{\beta \mid a_i \leq \beta_i \leq b_i, i = 1, \ldots, k\}$;

2) $\Theta = \{\theta \mid \theta' A\theta \leq b\}$, where A is a known $k \times k$ positive definite matrix and $b > 0$ is a known scalar. Here they transform the problem by supposing, without loss of generality, that A is symmetric so that $A = P'DP$ with D diagonal and P orthogonal. Then, with $V = ZP'$ and $\beta = P\theta$, the model becomes $X = V\beta + \varepsilon$ with β restricted to $C = \{\beta \mid \beta'D\beta \leq b\}$.

They then apply Moors (1981, 1985) to estimate β using squared-error loss, i.e., they find, for each x, a convex subset C_x of C with the properties that

$$P_\beta(\delta(X) \text{ not in } C_X) > 0 \text{ for some } \beta \in C \implies \delta(X) \text{ is inadmissible}$$

and the projection of $\delta(X)$ unto C_X dominates $\delta(X)$. They also show that C_x is the convex closure of the range of $h_x(\beta)$, $\beta \in C$, where $h_x(\beta) = \beta \tanh(x'V\beta)$.

When $k = 1$ both problems reduce to the case where $X \sim \mathcal{N}(\beta, \sigma^2)$ with β restricted to an interval. The Moors–van Houwelingen results are then identical to those of Moors (1981, 1985) presented in Chapter 3, Section 3.2.

6.4 Discussion and open problems

None of the above authors touches upon the question of whether their dominators (and other estimators) satisfy (2.3), but the following can be said about it.

In the first place, all equality- and inequality-restricted estimators do satisfy (2.3). Further the Moors–van Houwelingen dominators are projections onto Θ and thus satisfy it. But, of the Judge–Yancey–Bock–Bohrer dominators (6.4)–(6.6), only (6.5) satisfies (2.3). This means that the dominators (6.4) and (6.6) of $\hat{\beta}$ can be improved upon by projection onto $\{\beta \mid \beta \geq 0\}$. Another dominator which does not satisfy (2.3) is the Ohtani–Wan (1998) dominator. These authors start with the Judge–Yancy–Bock–Bohrer dominator (6.4) and replace σ^2 by an estimator of it. Another question not mentioned by any of the above authors is concerned with their loss functions. All, except Wan (1994b), use squared-error loss, but those who transform the model (6.1) do not say anything about what this loss function implies for estimating θ. For the Judge–Yancey transformation the relationship between β and θ is given by $\beta = Q'(Z'Z)^{1/2}\theta$ so that

$$\beta'\beta = \theta'(Z'Z)^{1/2}QQ'(Z'Z)^{1/2}\theta = \theta'Z'Z\theta.$$

So their loss function for estimating θ becomes $(d - \theta)'Z'Z(d - \theta)$.

Escobar and Skarpness (1987) and Wan (1994a) have $r_1 \leq b'\theta \leq r_2$ and take $\beta = B\theta$ with $B = (b, D)$ for a non-singular $k \times (k - 1)$ matrix D. So here $\beta'\beta = \theta'B'B\theta$.

For the Moors–van Houwelingen results with $\Theta = \{\theta \mid a \leq A\theta \leq b\}$, the transformation gives $\beta = A\theta$ so that $\beta'\beta = \theta'A'A\theta$, while for the case where $\Theta = \{\theta \mid \theta'B\theta \leq b\}$ they have $\beta = P\theta$ with $B = P'DP$ with D diagonal and P orthogonal. So here $\beta'\beta = \theta'P'P\theta = \theta'\theta$.

Finally, comparisons between the various dominators are not mentioned in any of the above-quoted papers. For instance, how do the Moors–van Houwelingen dominators for the case where $\beta_1 \geq 0$ compare with the Judge–Yancey–Bock–Bohrer dominator (6.5)?

7

Other properties

In this chapter several properties of restricted-parameter-space estimators which have not, or have barely, been touched upon up to now are presented. In Section 7.1 ideas of and results in weighted likelihood inference are applied to two restricted-normal-mean problems. Section 7.2 contains some (more) results, mostly numerical, on how robust restricted-parameter-space results are to misspecification of Θ; on how much one gains in risk function from restricting the parameter space; and, for one case, on how much bias this introduces.

7.1 Ideas from weighted likelihood estimation

As mentioned in Chapter 5, Section 5.1, van Eeden and Zidek (2002, 2004) study a class of estimators of μ_1 when $X_i \sim^{ind} \mathcal{N}(\mu_i, \nu_i^2)$, $i = 1, 2$, with known ν_i^2's for the case where $\mu_1 \leq \mu_2$ as well as for the case where $|\mu_2 - \mu_1| \leq c$ for a known $c > 0$. In constructing these two classes of estimators the authors use ideas and results from weighted likelihood. These ideas are the following: suppose Π_1 is the population of interest and data, X, say, from that population are available. Suppose that, in addition to the data X one has data, Y, say, from a population Π_2 which is "close" to Π_1. Then Y might well contain information which could be used to improve the precision of inference procedures based on X concerning Π_1.

Such data sets occur, e.g., in small-area estimation problems where one might have data from areas adjoining the area of interest. In such a case, by judiciously using Y together with X, one could "borrow strength from one's neighbours". Another example of such data occurs in regression analysis. Smoothing procedures, e.g., use data at nearby values of the independent variable(s) to improve on the estimate of the regression curve. In the case of the normal-mean problems of van Eeden and Zidek, when $|\mu_2 - \mu_1| \leq c$ for a small c, one should be able to use X_2 together with X_1 to improve on the best estimator of μ_1 based on X_1 alone. But, of course, by using Y one introduces bias in one's estimators and one has to decide on how much bias one wants to trade for how much increase in precision.

So, how does one go about using Y together with X for inference procedures concerning Π_1? In their 2002 and 2004 papers, van Eeden and Zidek use F. Hu's (1994, 1997) and Hu and Zidek's (2001, 2002) (relevance-)weighted likelihood method, which can be described as follows. Let independent observations $X_{i,j}, j = 1, \ldots, n_i$ from a populations Π_i be available, $i = 1, \ldots, k$, and suppose that Π_1 is the population of interest. Let f_i and F_i be the density and distribution function of $X_{i,j}, j = 1, \ldots, n_i, i = 1, \ldots, k$, and suppose a predictive density, g, say, of $X_{1,1}$ must be found to maximize $\int \log g(x) dF_1(x)$, i.e. the authors use Akaike's entropy maximization principle. This maximization is done under the restriction that Π_1 is "close" to $\Pi_2, \ldots, \Pi_k$, i.e., under the restriction that $\int \log g(x) dF_i(x) > c_i, i = 2, \ldots, k$ for given $c_2, \ldots, c_k$. In the parametric case where $f_i(x) = f_i(x \mid \theta_i)$ this leads to finding $\lambda_{i,j}$ maximizing

$$\prod_{i=1}^{k} \prod_{j=1}^{n_i} f_i^{\lambda_{i,j}/n_i}(x_{i,j} \mid \theta_i).$$

Hu (1994) in his relevance-weighted likelihood requires $\lambda_{i,j} \geq 0$, but van Eeden and Zidek (2002, 2004) noticed that many known estimators can be derived from weighted likelihood (WL) by allowing negative weights. Wang (2001), in his PhD thesis, dropped the non-negativity condition when looking for optimum weights.

Applying WL to the case of $X_i \sim^{ind} \mathcal{N}(\mu_i, \nu_i^2), i = 1, 2$, with known ν_i^2's we have $k = 2$ and $n_1 = n_2 = 1$. Putting $\lambda_{i,1} = \lambda_i, i = 1, 2$, the weighted likelihood estimator (WLE) of μ_1 can be shown to be given by $\delta_{WLE}(X_1, X_2) = X_1 + Z\alpha$, where $Z = X_2 - X_1$ and α is obtained from the weights by

$$\frac{\lambda_2}{\lambda_1} = \frac{\nu_2^2}{\nu_1^2} \frac{\alpha}{1 - \alpha}.$$

The maximization criterion here leads to minimization of the MSE and the optimal choice, in case $\Delta = \mu_2 - \mu_1$ is known, is given by

$$\alpha_{opt} = \frac{\nu_1^2}{\nu_1^2 + \nu_2^2 + \Delta^2}.$$

Since Δ is unknown, it must be estimated and van Eeden and Zidek (2002, 2004) propose to replace Δ by its MLE, so that the adaptively weighted likelihood estimators of μ_1 become (see (5.7))

$$\delta_{WLE}(X_1, X_2) = X_1 + Z \frac{\nu_1^2}{\nu_1^2 + \nu_2^2 + (\max(0, Z))^2}$$

when $\mu_1 \leq \mu_2$ and

$$\delta_{WLE}(X_1, X_2) = X_1 + Z \frac{\nu_1^2}{\nu_1^2 + \nu_2^2 + (\min(Z^2, c^2))}$$

when $|\mu_2 - \mu_1| \leq c$.

As seen in Chapter 5, Section 5.1, van Eeden and Zidek (2002, 2004) do not study only δ_{WLE}, but a class of estimators of μ_1 of the form $X_1 + \varphi(Z)$. Two examples of members of that class are its MLE (see (5.5)) and its Pitman estimator (see (5.6)). Other examples studied by van Eeden and Zidek are, for $\mu_1 \leq \mu_2$,

$$\delta_{\min}(X_1, X_2) = X_1 + \min(0, Z)$$

and, for $|\mu_2 - \mu_1| \leq c$,

$$\delta_m(X_1, X_2) = X_1 + Z \frac{\nu_1^2}{\nu_1^2 + \nu_2^2 + c^2}$$

and

$$\delta_M(X_1, X_2) = X_1 + Z \frac{\nu_1^2}{\nu_1^2 + \nu_2^2 + \min(Z^2, c^{*2})}$$

where $c^* = c \tanh(c|Z|/(\nu_1^2 + \nu_2^2))$. The estimator δ_m is a kind of minimax estimator. It maximizes (over α) the minimum (over $\Delta^2 \leq c^2$) improvement in MSE of $X_1 + \alpha Z = (1-\alpha)X_1 + \alpha X_2$ over X_1. The estimator δ_M is obtained by replacing, in δ_{WLE}, the MLE of Δ by Moors' dominator of it (see Chapter 3, Section 3.2).

Formulas for the MSE's of these estimators, as well as graphs of their biases and MSE's can be found in van Eeden and Zidek (2002, 2004).

7.2 Robustness, gains in risk function and bias

In this section robustness results with respect to missspecification of the parameter space are presented. These results are of the form: δ_1 dominates δ_2 not only on Ω but on the larger space $\Omega' \supset \Omega$. Some results of this kind are mentioned in Chapter 6, where the linear model $X = Z\theta + \varepsilon$ is considered with X an $n \times 1$ vector of observables, Z an $n \times k$ nonstochastic matrix, ε an $n \times 1$ vector of random variables with $\mathcal{E}\varepsilon = 0$ and $cov(\varepsilon) = \Sigma$ and squared-error loss. For $\varepsilon \sim \mathcal{N}_n(0, \sigma^2 I)$, as seen in Chapter 6, Section 6.1, Thomson and Schmidt (1982) show that, when $\Theta = \{\theta \mid \theta_1 \geq 0\}$, the MLE of θ_i dominates its unrestricted MLE δ_i^* on $\Theta' = \{\theta \mid \theta_1 \geq -.84\sigma_1\}$, where σ_1^2 is the variance of δ_1^*, and this result holds for all $i = 1, \ldots, k$.

Still for the linear model, Escobar and Skarpness (1987) (for the notation see Chapter 6, Section 6.2) compare, for squared-error loss, the restricted MLE $\hat{\beta}$ of β with its unrestricted one δ in the model $X = V\beta + \varepsilon$ when $r_1 \leq \beta_1 \leq r_2$ and $\varepsilon \sim \mathcal{N}_n(0, \sigma^2 I)$. As already noted there, they show that, when $\beta_1 \in [r_1, r_2]$, $\hat{\beta}_1$ dominates δ_1. But they also show that for $i = 1$ this domination holds on a larger interval $I \supset [r_1, r_2]$. Their proof goes as follows. They first show that

$$\Delta(\beta_1) = \frac{MSE_{\beta_1}(\delta_1) - MSE_{\beta_1}(\hat{\beta}_1)}{\sigma_1^2}$$

$$= -s_o\phi(s_o) + (1 - s_o^2)\Phi(s_o) + s_1\phi(s_1) + (1 - s_1^2)\Phi(-s_1),$$

where $s_o = (r_1 - \beta_1)/\sigma_1$, $s_1 = (r_2 - \beta_1)/\sigma_1$, σ_1^2 is the variance of δ_1 and ϕ and Φ are, respectively, the density and distribution function of the standard normal distribution. Using the fact that, for $r_1 \leq \beta_1 \leq r_2$, $s_o \leq 0$ and $s_1 \geq 0$ and the fact that $\phi(c) > c(1 - \Phi(c))$ for $c \geq 0$, it follows that $\Delta(\beta_1) > 0$ on $[r_1, r_2]$ and thus > 0 on a larger interval $I \supset [r_1, r_2]$. The authors give graphs of $\Delta(\beta_1)$ as a function of β_1 for two values of $d = s_1 - s_o$ and describe their features. But Ohtani (1987) looks at the same problem. He notes that one may, without loss of generality, suppose that $r_1 = -r_2$ and he gives graphs of $1 - \Delta(\beta_1)$ for five values of d. Both sets of authors note that $1 - \Delta(\beta_1) = MSE_{\beta_1}\hat{\beta}_1/\sigma_1^2$ has, for $\beta_1 \in [-r_2, r_2]$, its minimum at $\beta_1 = 0$ when d is small and its maximum at $\beta_1 = 0$ when d is large. From Ohtani's graphs it seems, as he notes, that $d \approx 1.5$ is the "turning" point. As an example of these author's robustness results: for $d = 1$, $\hat{\beta}_1$ dominates δ_1 for $\beta_1/\sigma_1 \in \approx (-1.8, 1.8)$, while for $d = 3$, $\hat{\beta}_1$ dominates δ_1 for $\beta_1/\sigma_1 \in \approx (-2.25, 2.25)$.

Similar results for the linear model, mostly based on graphical comparisons of risk functions, can (as mentioned in Chapter 6) be found in Thomson(1982), Judge, Yancey, Bock and Bohrer (1984) and Wan (1994a,b).

Two other examples of this kind of robustness can be found in van Eeden and Zidek (2002, 2004). As already mentioned in Chapter 5, Section 5.1, these authors consider several estimators of μ_1 when $X_i \sim^{ind} \mathcal{N}(\mu_i, \nu_i^2)$, $i = 1, 2$ with known ν_i^2's and either $\mu_1 \leq \mu_2$ or $|\mu_2 - \mu_1| \leq c$ for a known $c > 0$ with squared-error loss. For each of these two cases they give dominators for some of their inadmissible estimators as well as formulas for and graphs of the risk functions of these estimators. These results show, for the case where $|\mu_2 - \mu_1| \leq c$, that the domination results hold strictly on the whole interval $[-c, c]$ and thus on a larger interval. For the case with $\Theta = \{\mu \mid \mu_1 \leq \mu_2\}$, the Pitman estimator $\delta_P(X_1, X_2)$ and the estimator $\delta_{\min}(X_1, X_2) = \min(X_1, X_2)$ dominate X_1. However, their MSE's are equal to ν_1^2 when $\Delta = \mu_2 - \mu_1 = 0$ and each MSE has a strictly negative derivative (with respect to Δ) at $\Delta = 0$ implying that these domination results do not hold for $\Delta < 0$. But for this case the authors also show that the MLE strictly dominates X_1 on $[0, \infty)$ so that for that case the domination holds on an interval $(-C, \infty)$ for some $C > 0$.

In their 2004 paper van Eeden and Zidek also propose and study robust Bayes estimators. Given that c might not be known exactly, that a small c is desirable for maximizing the benefit of a restricted space, but that a c too small (so that in fact $|\mu_2 - \mu_1|$ might be larger than c), would lead to an underreporting of the risk of the estimator, they propose the following hierarchical Bayes estimator. At the first stage of the analysis assume that $\mu_i \sim^{ind} \mathcal{N}(\xi_i, \gamma_i^2)$, $i = 1, 2$ and at the second stage suppose that $|\xi_2 - \xi_1| \leq c$. Then, conditionally on (X_1, X_2), $\mu_i \sim^{ind} \mathcal{N}(\eta_i X_i + (1 - \eta_i)\xi_i, \nu_i^2 \gamma_i^2 / \lambda_i^2)$, $i = 1, 2$, where $\lambda_i^2 = \nu_i^2 + \gamma_i^2$ and $\eta_i = \gamma_i^2 / \lambda_i^2$ and, for an estimator $\hat{\mu}_1(X_1, X_2)$ of μ_1,

$$
\left.
\begin{aligned}
& \mathcal{E}\left((\hat{\mu}_1(X_1, X_2) - \mu_1)^2 | X_1, X_2\right) = \\[1em]
& (\hat{\mu}_1(X_1, X_2) - (\eta_1 X_1 + (1 - \eta_1)\xi_1))^2 + \frac{\nu_1^2 \gamma_1^2}{\lambda_1^2}.
\end{aligned}
\right\}
\tag{7.1}
$$

(There is a misprint in the authors' formula (6) for this MSE.)

Now use the fact that, marginally, $X_i \sim^{ind} \mathcal{N}(\xi_i, \lambda_i^2)$ and estimate ξ_1 by its Pitman estimator under the restriction $|\xi_2 - \xi_1| \leq c$ given by (see the second line of (5.6))

$$
X_1 + \frac{\lambda_1^2}{\lambda} \frac{\phi((Z - c)/\lambda) - \phi((Z + c)/\lambda)}{\Phi((Z + c)/\lambda) - \Phi((Z - c)/\lambda)},
$$

where $\lambda^2 = \lambda_1^2 + \lambda_2^2$ and $Z = X_2 - X_1$. Then the robust Bayes estimator, obtained by substituting this Pitman estimator of ξ_1 into (see (7.1)) $\eta_1 X_1 + (1 - \eta_1)\xi_1$, becomes

$$
\delta_{rb}(X_1, X_2) = X_1 + \frac{\nu_1^2}{\lambda} \frac{\phi((Z - c)/\lambda) - \phi((Z + c)/\lambda)}{\Phi((Z + c)/\lambda) - \Phi((Z - c)/\lambda)}.
$$

Taking $\gamma_1^2 = \gamma_2^2 = .35$ in δ_{rb}, the authors compare it, graphically, with the Pitman estimator δ_P as well as with one of their other estimators of μ_1, namely,

$$
\delta_{WLE}(X) = X_1 + Z \frac{\nu_1^2}{\nu_1^2 + \nu_2^2 + \min(Z^2, c^2)}.
$$

Their Figure 9 shows that this δ_{rb} and δ_{WLE} have almost identical risk functions and their Figure 8 then shows that this δ_{rb} is more robust to misspecification of c than is δ_P.

Of course, these robustness results are not all that surprising because: suppose δ_1 dominates δ_2 on Θ and suppose their risk functions are continuous on $\Theta' \supset \Theta$, then if $R(\delta_1, \theta) < R(\delta_2, \theta)$ at a boundary point, θ_o, say, of Θ which is not a boundary point of Θ', then there is an open subset of Θ' containing

θ_o on which with $R(\delta_1, \theta) - R(\delta_2, \theta) < 0$.

For the estimation of a positive normal mean based on $X \sim \mathcal{N}(\theta, 1)$, Katz (1961) showed admissibility and minimaxity of the Pitman estimator δ_K for squared-error loss (see Chapter 4, Section 4.3). Maruyama and Iwasaki (2005) study the robustness properties of these results to misspecification of the variance σ^2 of X. They show that, when $\sigma^2 \neq 1$, δ_K is minimax if and only if $\sigma^2 > 1$. Concerning the admissibility when $\sigma^2 \neq 1$ they show that δ_K is admissible if and only if σ^{-2} is a positive integer. This is an example of extreme non-robustness. No matter how close $\sigma^2 < 1$ is to 1, δ_K is not minimax. And admissibility happens "almost never".

On questions of how much can be gained, risk-function-wise, several results have already been mentioned in earlier chapters:

1) In Chapter 3, for the case of a lower-bounded normal mean as well as for the case of a symmetricaly restricted binomial parameter, numerical evidence is presented on by how much known dominators of the MLE lower its risk function;
2) In Chapter 4, numerical as well as theoretical evidence is given on by how much restricting the parameter space can lower the minimax value for the problem;
3) In Chapter 6 graphs are mentioned in which various estimators of θ are compared for the model $X = Z\theta + \varepsilon$ with restrictions on θ.

More numerical and graphical comparisons between estimators for the restricted problem as well as between "best" ones for the unrestricted case and "good" ones for the restricted case are presented in this section.

Some early results are those of Mudholkar, Subbaiah and George (1977). They consider the MLEs of two Poisson means μ_1 and μ_2 with $\mu_1 \leq \mu_2$ based on independent X_1, X_2. For the MLE $\hat{\mu}_1$ of μ_1 they show that its bias $B_{MLE,1}$ of is given by

$$B_{MLE,1} = \frac{-\mu_1(1 - F(2\mu_2; 2, 2\mu_1)) + \mu_2 F(2\mu_1; 4, 2\mu_1)}{2}, \qquad (7.2)$$

where $F(a; b, c)$ is the distribution function of a non-central χ^2 random variable with b degrees of freedom and non-centrality parameter c, evaluated at a. For the second moment of $\hat{\mu}_1$ they find

$$4\mathcal{E}_\mu \hat{\mu}_1^2 = \mu_1^2 + \mu_1 + 3\mu_1^2 F(2\mu_2; 4, 2\mu_1) + 3\mu_1 F(2\mu_2; 2, 2\mu_1)$$

$$+ \mu_2^2 F(2\mu_1; 6, 2\mu_2) + \mu_2 F(2\mu_1; 4, 2\mu_2) + 2\mu_1\mu_2 F(2\mu_1; 2, 2\mu_2)$$

and they use (7.2) to show that (as is intuitively obvious) $B_{MLE,1} < 0$ and, for each $\mu_1 > 0$, converges to zero as $\mu_2 \to \infty$. They have similar formulas for the bias and second moment of $\hat{\mu}_2$ and, for $i = 1$ as well as for $i = 2$, they give numerical values of $B_{MLE,i}$ and $\mathcal{E}_\mu(\hat{\mu}_i - \mu_i)^2$ for $\mu_2 = 10(10)50$ and, for each μ_2, for five values of $\mu_1 \leq \mu_2$. An example of these results is given in Tab 7.1. This table gives the MSE of $\hat{\mu}_2$ when $\mu_2 = 20$ and we see that $\text{MSE}(\hat{\mu}_2) < \mu_2 = \text{MSE}(X_2)$ for all μ_1.

Table 7.1. $X_i \sim^{ind} \text{Poisson}(\mu_i)$, $i = 1, 2$, $\mu_1 \leq \mu_2 = 20$.

μ_1	8	10	12	16	20
$\text{MSE}(\hat{\mu}_2)$	19.86	19.55	18.92	16.83	15.63

All their other numerical results for the MSE's show the same pattern: $\text{MSE}(\hat{\mu}_i) < \mu_i = \text{MSE}(X_i)$ for $i = 1, 2$ for all the values of $\mu_1 \leq \mu_2$ for which they present results. From these results (combined with results on the bias of the estimators) they conclude (their p. 93): "$\hat{\mu}_1$ and $\hat{\mu}_2$ are biased but their MSE's are less than respectively μ_1 and μ_2, the MSE's of X_1 and X_2." But, as we saw in Chapter 5, Section 5.1, Kushary and Cohen (1991) show that, if $\delta(X_2)$ is, for squared-error loss, admissible for estimating μ_2 based on X_2 alone, then it is admissible for estimating μ_2 based on (X_1, X_2). So, the question is: "For what values of (μ_1, μ_2) with $\mu_1 \leq \mu_2$ is $\text{MSE}(X_2) = \mu_2 < \text{MSE}(\hat{\mu}_2)$?". The answer is in the following lemma. My thanks to Arthur Cohen for pointing out that the answer should be : "For small values of μ_1.".

Lemma 7.1 *For* $X_i \sim^{ind} \text{Poisson}(\mu_i)$, $i = 1, 2$, *with* $\mu_1 \leq \mu_2$,

$$MSE(\hat{\mu}_2) > \mu_2 \ for \ 0 < \mu_1 \leq \mu_2 < .25,$$

where $\hat{\mu}_2 = \max(X_2, (X_1 + X_2)/2)$ *is the MLE of* μ_2.

Proof. The MSE of $\hat{\mu}_2$ is given by

$$\text{MSE}(\hat{\mu}_2) = \sum_{i=0}^{\infty} \sum_{j=0}^{\infty} \left(\max\left(\frac{i+j}{2}, j\right) - \mu_2 \right)^2 \frac{e^{-\mu_1}\mu_1^i}{i!} \frac{e^{-\mu_2}\mu_2^j}{j!}.$$

So, for each μ_2,

$$\lim_{\mu_1 \to 0} \text{MSE}(\hat{\mu}_2) = \mu_2. \tag{7.3}$$

Further

$$\frac{d}{d\mu_1} \text{MSE}\,(\hat{\mu}_2) =$$

$$\left. \begin{aligned} &-\sum_{j=0}^{\infty}\left(\max\left(\frac{j}{2},j\right)-\mu_2\right)^2 \frac{e^{-\mu_2}\mu_2^j}{j!}e^{-\mu_1}+\\ &\sum_{i=1}^{\infty}\sum_{j=0}^{\infty}\left(\max\left(\frac{i+j}{2},j\right)-\mu_2\right)^2\times\\ &\left(\frac{e^{-\mu_1}\mu_1^{i-1}}{(i-1)!}-\frac{e^{-\mu_1}\mu_1^i}{i!}\right)\frac{e^{-\mu_2}\mu_2^j}{j!}. \end{aligned} \right\} \tag{7.4}$$

From (7.4) it follows that

$$\frac{d}{d\mu_1}\text{MSE}\,(\hat{\mu}_2)|_{\mu_1=0} = -\mu_2 + \sum_{j=0}^{\infty}\left(\max\left(\frac{1+j}{2},j\right)-\mu_2\right)^2\frac{e^{-\mu_2}\mu_2^j}{j!}$$

$$= -\mu_2 + \left(\frac{1}{2}-\mu_2\right)^2 e^{-\mu_2} + \sum_{j=1}^{\infty}(j-\mu_2)^2\frac{e^{-\mu_2}\mu_2^j}{j!}$$

$$= \left(\frac{1}{2}-\mu_2\right)^2 e^{-\mu_2} - \mu_2^2 e^{-\mu_2} = \left(\frac{1}{4}-\mu_2\right)e^{-\mu_2},$$

which, together with (7.3), proves the result.   ♡

This result shows how careful one should be with drawing conclusions from numerical results.

How much one gains risk-function-wise and how much bias is introduced when restricting a parameter space: it rather heavily depends on the model and, given a model, it rather heavily depends on the estimand. But, of course, any improvement result gives only a lower bound on the possible improvements. Here are some examples (note that many of the quoted numerical values are obtained from graphs or from ratios of numerical (Monte-Carlo) results and then are only approximate values):

For the van Eeden and Zidek (2002) normal-mean problem with $k=2$ and $\mu_1 \le \mu_2$, their estimator of μ_1 given by (see (5.7))

$$\delta_{WLE}(X_1,X_2) = X_1 + \frac{\nu_1^2(X_2-X_1)}{\nu_1^2+\nu_2^2+(\max(0,(X_2-X_1)))^2},$$

gives, when $\nu_1 = \nu_2 = 1$, an approximately 20–40% decrease in MSE over X_1 when $\Delta = \mu_2 - \mu_1 \le .1$, a larger decrease than any of their other three estimators, the MLE, $\delta_{\min}(X_1,X_2) = \min(X_1,X_2)$ and the Pitman estimator $\delta_P(X_1,X_2)$, give in that same interval. Still for $\nu_1 = \nu_2 = 1$, the estimator δ_{WLE} also has (see their Figure 1) the smallest absolute bias among these estimators in the same interval. But, for larger Δ ($\Delta > 2$, or so), its MSE is > 1, the MSE of X_1 and its bias there is the largest among the four estimators. The estimators $\delta_{\min}(X_1, X_2)$ and δ_P have their maximum gain in risk function ($\approx 20\%$) for Δ in the interval $(1, 2)$, but these estimators do not gain anything at $\Delta = 0$. For all $\Delta \geq 0$ their absolute biases are the largest for those among the four estimators. The MLE might look like a good compromise: not too badly biased relative to the other three and a 25% gain in MSE at $\Delta = 0$. But, the MLE is inadmissible. The authors do give a class of dominators for each of their inadmissible estimators, but no graphs comparing inadmissible estimators with dominators. Whether δ_{WLE} is admisible is unknown, but it is a smooth estimator, which the MLE is not. And it dominates the MLE for $0 \leq \Delta \leq A$, where A is slightly less than 2. The 2002 paper of van Eeden and Zidek also contains graphs of the MSE's of their estimators for cases with $\nu_1^2 \neq \nu_2^2$.

Some of the above-mentioned properties of δ_P, $\delta_{\min}$ and the MLE for the ordered-normal-mean problem with $\nu_1^2 = \nu_2^2 = 1$ can also be obtained from numerical results of Al-Saleh (1997) and Iliopoulos (2000). Al-Saleh gives numerical values for the MSE of δ_P as well as of $\delta_{\min}$ for $\Delta = 0(.2)4(1)6$ and 10 and for the MLE such values (in the form of % decrease in MSE over X_1) can be found in the last line of Iliopoulos' (2000) Table 1. Iliopoulos (see Chapter 5, Section 5.1) has results for the normal-mean problem with $k = 3$ and $\mu_1 \leq \mu_2 \leq \mu_3$. He estimates μ_2 and taking $\mu_1 = -\infty$ in his results gives the MLE of the smaller one of two ordered normal means. Iliopoulos (2000) also has numerical results for comparing his estimator with the middle one of his three ordered normal means by the linex loss function. Tab 7.2 contains some of the numerical values obtained by Al-Saleh (1997) and Iliopoulos (2000).

Table 7.2. % MSE improvement over X_1, normal means.

$\mu_2 - \mu_1 \geq 0$	0	.20	.40	.50	.60	.80	1.00	1.40	1.60	2.00	4.00
$\delta_{\min}$	0	9	15	-	19	20	20	17	15	10	0
δ_P	0	6	11	-	14	17	19	21	21	10	6
MLE	25.0	-	-	22.5	-	-	17	-	-	6.5	.2

Al-Saleh (1997) also has numerical results for the bias of δ_P and $\delta_{\min}$. His results are summarized in Tab 7.3. A formula for the bias of the MLE of μ_1 can be obtained from van Eeden and Zidek's (2002) Lemma A.3, while Sampson, Singh and Whitaker (2003) give formulas for the biases of the MLEs of μ_1 and μ_2.

Table 7.3. Biases, normal means, estimate μ_1.

$\mu_2 - \mu_1 \geq 0$	0	.20	.40	.60	.80	1.00	1.40	1.60	2.00	4.00
$\delta_{\min}$	.56	.47	.39	.31	.25	.20	.12	.09	.05	.00
δ_P	.66	.58	.53	.47	.42	.38	.30	.26	.20	.03

From Tab 7.2 and Tab 7.3, as well as from van Eeden and Zidek (2002), one sees that neither one of $\delta_{\min}$ and δ_P is very good as an estimator of μ_1 when Δ is close to zero: they do not improve much over X_1 and both are very biased.

For the normal-mean problem with $k = 2$, $\nu_1^2 = \nu_2^2 = 1$ and $\Delta = |\mu_2 - \mu_1| \leq c$, van Eeden and Zidek (2004) give graphs of the MSE of the four estimators defined in Section 7.1 for $c = 1$ and $-2 \leq \Delta \leq 2$. Each of these estimators gives an MSE reduction (relative to X_1) of more than 40% in the middle of the interval $(-c, c)$, while over the whole interval $(-c, c)$ this reduction is at least 30%. The authors do not give any bias results for this case, but for several inadmissible estimators and their dominators they give graphs of their MSEs. For the MLE for instance, which has an MSE which is almost constant at about .67 on most of the interval $(-c, c)$, this dominator's MSE varies from about .57 at $\Delta = 0$ to about .67 at the endpoints of the interval $(-c, c)$. However, the dominator is much less robust to misspecification of Θ than is the MLE. As in their 2002 paper, van Eeden and Zidek (2004) also give graphs of MSE's of their estimators for cases where the variances are not equal.

Vijayasree, Misra and Singh (1995) present, for squared-error loss, several tables with numerical comparisons of risk functions for pairs of estimators of location or scale parameters of exponential distributions when $k = 2$ (see Chapter 5, Sections 5.1 – 5.3). In their Table 1, e.g., they compare, for estimating the smaller scale parameter when the location parameters are known, their dominator

$$\begin{cases} \dfrac{X_1}{n_1 + 1} & \text{when } \dfrac{X_1}{n_1 + 1} \leq \dfrac{X_2}{n_2} \\[3mm] \dfrac{X_1 + X_2}{n_1 + n_2 + 1} & \text{when } \dfrac{X_1}{n_1 + 1} > \dfrac{X_2}{n_2} \end{cases}$$

of $X_1/(n_1 + 1)$ with their dominator

$$\begin{cases} \dfrac{X_1}{n_1} & \text{when } \dfrac{X_1}{n_1} \leq \dfrac{X_2}{n_1 + 1} \\[3mm] \dfrac{X_1 + X_2}{n_1 + n_2 + 1} & \text{when } \dfrac{X_1}{n_1} > \dfrac{X_2}{n_1 + 1} \end{cases}$$

of the MLE. Such results do not give any information on how much the dominators improve, risk-function-wise, on the dominated estimators. Nor do any of the other eight tables in their paper give such information.

Misra and Singh (1994), using squared-error loss, look at the estimation of two ordered exponential location parameters when the scale parameters are known (see Chapter 5, Section 5.1). For the special case where $\nu_2 n_1 = \nu_1 n_2$, their best (mixed) dominator $\delta_{1,\alpha}$ of the unrestricted MRE, $X_1 - \nu_1/n_1$, of the smaller location parameter has $\alpha = \alpha^* = .25$ as its mixing parameter, while their best mixed dominator $\delta_{2,\beta}$ of $X_2 - \nu_2/n_2$ as an estimator of the larger parameter has $\beta = \beta^* = .75$. The authors give the following formulas for the MSEs of these dominators

$$\text{MSE}_\Delta(\delta_{1,\alpha}) = 1 - (1 - \alpha)(\Delta + \alpha + 1/2)e^{-\Delta}$$

$$\text{MSE}_\Delta(\delta_{2,\beta}) = 1 - (1 - \beta)(\beta - 1/2)e^{-\Delta},$$

where $\Delta = \mu_2 - \mu_1$. Using these formulas and the author's Table 1 and Table 2 gives the percent risk improvements presented in Tab 7.4. The third line of the table gives the % risk improvements of $(\delta_{1,\alpha^*}, \delta_{2,\beta^*})$ over $(X_1 - \nu_1/n_1, X_2 - \nu_2/n_2)$ when the loss function is the sum of the squared errors. Given that, when $\nu_2 n_1 = \nu_1 n_2$, the MRE's of μ_1 and μ_2 have the same (constant) risk function, the numbers in the third line in Tab 7.4 are the averages of the corresponding ones in the first and second lines.

Table 7.4. % MSE improvement over the MRE, exponential location.

$\mu_2 - \mu_1 \geq 0$	0.00	0.05	0.10	0.20	0.25	0.50	0.75	1.00	3.00	5.00	6.00
δ_{1,α^*}	56	56	58	58	58	57	53	48	14	3	1
δ_{2,β^*}	6.2	5.9	5.7	5,1	4.9	3.8	3.0	2.3	0.0	0.0	0.00
$(\delta_{1,\alpha^*}, \delta_{2,\beta^*})$	31	31	32	32	32	30	28	25	7	2	1

So, for small values of Δ, δ_{1,α^*} substantially improves on $X_1 - \nu_1/n_1$. But δ_{2,β^*}'s improvements are very small. Moreover, as remarked on in Chapter 5, Section 5.1, $\delta_{2,\beta^*} > X_2$ with positive probability for all parameter values.

Jin and Pal (1991) also have results for estimating two exponential location parameters under restrictions on the parameter space (see Chapter 5, Section 5.3). Their scale parameters are unknown and they dominate, for squared-error loss, the vector $(X_1 - T_1/n_1^2, X_2 - T_2/n_2^2)$, where $T_i = \sum_{j=1}^{n_i}(X_{i,j} - X_i)$, $i = 1, 2$. They give Monte-Carlo estimators of percent improvements in risk function for each of their dominators for several values of the parameters and the sample sizes. Tabs 7.5 and 7.6 contain examples of these Monte-Carlo results for the case where the parameters are restricted by $\mu_1 \leq \mu_2$. Here their dominators are mixed estimators and the dominator of $X_2 - T_2/n_2^2$ is $> X_2$ with positive probability for all parameter values.

Table 7.5. % MSE improvement of the α-mixed estimator (5.14) of (μ_1, μ_2) over its MRE, ordered exponential location, $\mu_1 = 0.0$, $\mu_2 = 0.1$, $\nu_1 = 1.0$, $\nu_2 = 1.1$.

α	0.0	0.1	0.2	0.3	0.4	0.50
$n_1 = n_2 = 3$	12.82	17.72	21.32	23.64	24.67	24.67
$n_1 = 5, n_2 = 10$	22.64	27.69	31.18	33.11	33.47	33.47
$n_1 = n_2 = 10$	18.35	19.72	20.41	20.41	19.78	19.78

Table 7.6. % MSE improvement of the α-mixed estimator (5.14) of (μ_1, μ_2) over its MRE, ordered exponential location, $\mu_1 = 0.0$, $\mu_2 = 0.5$, $\nu_1 = 1.0$, $\nu_2 = 1.1$.

α	0.0	0.1	0.2	0.3	0.4	0.50
$n_1 = n_2 = 3$	18.13	18.55	18.56	18.16	17.34	17.34
$n_1 = 5, n_2 = 10$	15.81	15.41	14.80	13.99	12.97	12.97
$n_1 = n_2 = 10$	3.33	3.22	3.11	2.98	2.84	2.84

From these tables it is seen that the optimal α of the dominating mixed estimator depends upon the unknown parameters. One needs a larger α for a smaller difference $\Delta = \mu_2 - \mu_1$. But, intuitively, for a given α and a given Δ, the improvement should decrease with increasing sample sizes. It does in Tab 7.6, but it does not in Tab 7.5. Also, for a given α and given sample sizes, the improvement "should" decrease with Δ. It does for $n_1 = 5$, $n_2 = 10$ and for $n_1 = n_2 = 10$, but not for $n_1 = n_2 = 3$. Is something possibly wrong with the entries in the first line of Tab 7.5?

For estimating, with squared-error loss, ordered scale parameters $\mu_1 \leq \mu_2$ of exponential distributions, Vijayasree and Singh (1993) study (see Chapter 5, Section 5.2) a class of mixed estimators of each of the components of μ. They give numerical values for the efficiency of these estimators relative to the unrestricted MLE (UMLE). Tabs 7.7 and 7.8 give the percent improvements of these mixed estimators over the UMLE obtained from the authors' results. In both tables the results hold for equal sample sizes, $n_1 = n_2 = n$. Tab 7.7 gives these improvements for estimating μ_1 and the mixing parameter $\alpha_1 = n_1/(n_1 + n_2 + 1)$, while Tab 7.8 gives these improvements for estimating μ_2 and mixing parameter $\alpha^* = 1/2 - (1/2)^{2n} \binom{2n-1}{n}$.

Table 7.7. % MSE improvement over X_1/n_1 of the α_1-mixed estimator of μ_1 with $\alpha_1 = n_1/(n_1 + n_2 + 1)$, gamma scale.

$\mu_2/\mu_1 \geq 1$	1.0	1.3	1.5	2.0	3.0
$n_1 = n_2 = 5$	38	32	27	17	7
$n_1 = n_2 = 10$	34	25	18	8	1
$n_1 = n_2 = 50$	29	9	2	0	0

Table 7.8. % MSE improvement over X_2/n_2 of the α^*-mixed estimator of μ_2 with $\alpha^* = 1/2 - (1/2)^{2n} \binom{2n-1}{n}$, gamma scale.

$\mu_2/\mu_1 \geq 1$	1.0	1.3	1.5	2.0	3.0
$n_1 = n_2 = 5$	14	12	10	5	1
$n_1 = n_2 = 10$	17	13	9	3	0
$n_1 = n_2 = 50$	21	6	1	0	0

More numerical comparisons between estimators of ordered scale parameters of gamma distributions can (see Chapter 5, Section 5.2) be found in Misra, Choudhary, Dhariyal and Kundu (2002). They give MSEs for each of the components of the unrestricted MRE, of the MLE and of one of the estimators of Vijayasree, Misra and Singh (1995) as well as for their own estimators (5.11) and (5.12). They have tables for equal sample sizes $n_1 = n_2 = 1, 5, 10, 20$ and four pairs of unequal ones, each combined with ten values of μ_2/μ_1. Tab 7.9 (respectively, Tab 7.10) gives percent improvements of (5.11) (respectively, (5.12)) over the unrestricted MLE for estimating μ_1 (respectively μ_2) obtained from the Misra, Choudhary, Dhariyal and Kundu (2002) tables.

Comparing the results in Tab 7.9 (resp. Tab 7.10) with those in Tab 7.7 (resp. Tab 7.8), one sees that

Table 7.9. % MSE improvement over X_1/n_1 of (5.11), gamma scale.

$\mu_2/\mu_1 \geq 1$	1.00	1.25	1.67	2.00	2.50	3.33
$n_1 = n_2 = 5$	18	21	24	26	25	24
$n_1 = n_2 = 10$	9	18	20	20	18	14
$n_1 = n_2 = 20$	6	16	18	14	8	6

a) For estimating μ_1 with $\mu_2/\mu_1 \leq d$, the α_1-mixed estimator of Vijayasree and Singh is better than the Misra, Choudhary, Dhariyal and Kundu
   (2002) estimator, where d depends on the common sample size n. Further,
   as also seen in earlier tables, the improvement decreases with increasing
   n;
b) for estimating μ_2, the Misra, Choudhary, Dhariyal and Kundu (2002) estimator does much better than the α^*-mixed estimator of Vijayasree and
   Singh for μ_2/μ_1 not close to 1.

Table 7.10. % MSE improvement over X_2/n_2 of (5.12), gamma scale.

$\mu_2/\mu_1 \geq 1$	1.00	1.25	1.67	2.00	2.50	3.33
$n_1 = n_2 = 5$	16	34	40	38	34	29
$n_1 = n_2 = 10$	9	29	32	29	22	15
$n_1 = n_2 = 20$	4	26	22	18	10	6

Ghosh and Sarkar (1994) estimate (see Chapter 5, Section 5.3) the smaller one
of two ordered normal variances based on $Y_{i,j} \sim^{ind} N(\mu_i, \nu_i)$, $j = 1, \ldots, n_i$,
$i = 1, 2$, with $\nu_1 \leq \nu_2$. They give Monte-Carlo estimates of the percent decrease in MSE of their dominators of $X_1/(n_1 + 1)$, the MRE based on the
first sample alone. Tabs 7.11 and 7.12 contain these results for two of their
dominators, namely:

1) Their dominator T_4, which is the estimator (5.20) with $\phi(W)$ as (5.16). It
   dominates $(1 - \phi(W))X_1/(n_1 + 1)$ for this same ϕ and thus $X_1/(n_1 + 1)$.
   The value of ε used in this table is .02;
2) Their dominator T_8, which is the estimator (5.19) with ϕ as in (5.17). It
   dominates $(1 - \phi(V))X_1/(n_1 + 1)$ for this same ϕ and thus $X_1/(n_1 + 1)$.
   In this table $\varepsilon = .01$ is used.

Table 7.11. % MSE improvement over $X_1/(n_1 + 1)$ of (5.20) with ϕ as in (5.16), $\varepsilon = .02$ and $\mu_1 = 0$, normal variances.

$\nu_2/\nu_1 \geq 1$	1.00	1.25	1.50	2.00	2.50
$n_1 = n_2 = 5$	7.480	6.599	4.855	2.714	1.492
$n_1 = 10, n_2 = 5$	6.184	5.821	4.221	2.093	1.299
$n_1 = 5, n_2 = 10$	10.41	8.105	5.001	1.426	0.557
$n_1 = n_2 = 10$	10.13	7.726	4.824	1.273	0.248

Table 7.12. % MSE improvement over $X_1/(n_1 + 1)$ of (5.19) with ϕ as in (5.17), $\varepsilon = .01$ and $\mu_1 = 0$, normal variances.

$\nu_2/\nu_1 \geq 1$	1.00	1.25	1.50	2.00	2.50
$n_1 = n_2 = 5$	8.153	7.241	5.403	3.023	1.579
$n_1 = 10, n_2 = 5$	7.194	6.808	5.065	2.720	1.737
$n_1 = 5, n_2 = 10$	11.36	8.910	5.562	1.621	0.526
$n_1 = n_2 = 10$	11.60	8.748	5.618	1.652	0.421

Remark 7.1. There is a misprint in the Ghosh–Sarkar tables. They have ν_1/ν_2 instead of ν_2/ν_1 in the heading of the second column of their table.

From the complete tables of Ghosh and Sarkar one sees that (for the values of the parameters used in those table): of the nine estimators in Ghosh and Sarkar's tables, none dominates any of the other ones. But T_8 does, for $\nu_2/\nu_1 \leq$ 1.5, better than all the other ones. But, even for T_8, the improvements are small and, given that these are the best improvements possible (better than those for $\mu_1 \neq 0$), numerical results for other values of μ_1 would be a very helpful addition to the study of the properties of these dominators.

8

Existence of MLEs and algorithms to compute them

For the case where there are no nuisance parameters and Θ is determined by order restrictions among and bound restrictions on the parameters, this chapter gives, in Section 8.1, conditions for the existence of the MLE $\hat{\theta} = (\hat{\theta}_1, \ldots, \hat{\theta}_k)$ of $\theta = (\theta_1, \ldots, \theta_k)$ and, in Section 8.2, algorithms for computing it. The existence conditions are those of van Eeden (1956, 1957a,b, 1958) and we compare them with those of Brunk (1955), as well as with those of Robertson and Waltman (1968). The algorithms are those of van Eeden (1957a, 1958) and they are compared with some of the ones appearing in Barlow, Bartholomew, Bremner and Brunk (1972), in Robertson, Wright and Dykstra (1988) or in publications appearing after the Robertson–Wright–Dykstra book. Section 8.3 contains some results on norm-reducing properties of the MLE, while in Section 8.4 some algorithms for multivariate problems are presented.

The algorithms presented here do not apply to, e.g., star-shape-restricted θ_i. Algorithms for such Θ are described or referred to in Robertson, Wright and Dykstra (1988).

8.1 The model and the conditions

Let, for $i = 1, \ldots, k$, $X_{i,1}, \ldots, X_{i,n_i}$ be independent samples and let $f_i(x; \theta_i)$, $x \in R^1$, $\theta_i \in J_i$, be the density of $X_{i,1}$ with respect to a σ-finite measure ν where J_i is an interval and is the set of all θ_i for which $\int_{R^1} f_i(x; \theta_i) d\nu(x) = 1$. Suppose that we know that $\theta \in \Theta$, where

$$\Theta = \{\theta \mid \alpha_{i,j}(\theta_i - \theta_j) \le 0, 1 \le i < j \le k; \theta_i \in I_i, i = 1, \ldots, k\}. \quad (8.1)$$

Here $I_i = [a_i, b_i]$ with $a_i < b_i$, $i = 1, \ldots, k$, is a known subset of J_i. Further, the $\alpha_{i,j}$, $1 \le i < j \le k$, are known with $\alpha_{i,j} = 1$ if there exists an h such that $\alpha_{i,h} = \alpha_{h,j} = 1$. Otherwise $\alpha_{i,j}$ takes the value 0 or 1 and we suppose that Θ is not empty and $\ne \prod_{i=1}^k I_i$.

For $\theta_i \in I_i$, $i = 1, \ldots, k$, the log-likelihood function is given by

$$l(\theta) = \sum_{i=1}^{k} l_i(\theta_i), \tag{8.2}$$

where $l_i(\theta_i) = \sum_{j=1}^{n_i} \log f_i(x_{i,j}; \theta_i)$, $i = 1, \ldots, k$.

Suppose that l satisfies the following condition:

Condition A. For each $M \subset \{1, \ldots, k\}$ for which $I_M = \cap_{i \in M} I_i$ is not empty,

$$l_M(\theta) = \sum_{i \in M} l_i(\theta) \qquad \theta \in I_M \tag{8.3}$$

is strictly unimodal.

Here strictly unimodal means that there is a $v_M \in I_M$ such that, for $\theta \in I_M$, $l_M(\theta)$ is strictly increasing for $\theta < v_M$ and strictly decreasing for $\theta > v_M$. Note that v_M is the MLE of μ based on $X_{i,j}$, $j = 1, \ldots, n_i$, $i \in I_M$ under the condition that $\theta_i = \mu$ for all $i \in M$. Further note that the setfunction v_M, $M \subset \{1, \ldots, k\}$, satisfies the so-called Cauchy mean-value (CMV) property (see Robertson, Wright and Dykstra, 1988, p. 24), i.e., for each pair of subsets M_1 and M_2 of $\{1, \ldots, k\}$ with $M_1 \cap M_2 = \emptyset$, $\min(v_{M_1}, v_{M_2}) \le v_{M_1 \cup M_2} \le \max(v_{M_1}, v_{M_2})$. However, it does not necessarily have the strict CMV property which says that $v_{M_1 \cup M_2}$ is strictly between v_{M_1} and v_{M_2} (see Robertson, Wright and Dykstra, 1988, p. 390). An example of this non-strictness is the case where $X_{i,1}, \ldots, X_{i,n_i}$ are $\mathcal{U}(0, \theta_i)$, $i = 1, \ldots, k$. For this case $v_i = \max_{1 \le j \le n_i} X_{i,j}$, $i = 1, \ldots, k$, while for $M \subset \{1, \ldots, k\}$, $v_M = \max_{i \in M} v_i$.

A proof that Condition A is sufficient for the existence and uniqueness of the MLE of θ under the restriction $\theta \in \Theta$ can be found in van Eeden (1957a, 1958).

The above model is more general than the one considered by Brunk (1955). He supposes that the $X_{i,1}$, $i = 1, \ldots, k$, have an exponential-family distribution with density $f(x; \theta_i)$ and $\theta_i = \mathcal{E}_{\theta_i} X_{i,j}$. Further, his $I_i = J$, the natural parameter space of that family. He shows that the MLE of θ exists and is unique and gives explicit formulas, the so-called max-min formulas (see Section 8.2) for it. It is easily verified that Condition A is satisfied under Brunk's (1955) conditions and that in his case v_M has the strict CMV property.

An example where Brunk's (1955) conditions are not satisfied but Condition A is, is the case where the $X_{i,1}$ are $\mathcal{U}(0, \theta_i)$. Robertson and Waltman (1968) suppose, among other things, that the likelihoood is unimodal, but not strictly unimodal, i.e., the mode is not unique. An example where their conditions are satisfied, but neither Brunk's (1955) nor van Eeden's (1957a, 1958) are, is the double-exponential distribution. On the other hand, Robertson and Waltman's (1968) conditions are not satified for the above-mentioned uniform case.

Ayer, Brunk, Ewing, Reid and Silverman (1955) consider the special case where $X_{i,j} \sim \text{Bin}(1, \theta_i)$ and the θ_i are simply ordered, i.e., they satisfy $\theta_1 \leq \ldots \leq \theta_k$. They give an algorithm for computing the MLE of θ, which later came to be known as the PAVA (pool-adjacent-violators-algorithm). For the binomial case it says that $\hat{\theta}_i = \hat{\theta}_{i+1}$ when $X_i/n_i > X_{i+1}/n_{i+1}$, where $X_i = \sum_{j=1}^{n_i} X_{i,j}$. This reduces the k-dimensional problem to a (k - 1)-dimensional one and repeated application of the algorithm will give $\hat{\theta}$.

8.2 Algorithms

In this section we give, for the model as described in Section 8.1 and assuming that Condition A is satisfied, several algorithms for computing the MLE of θ and make some comparisons among algorithms.

We start with four theorems of van Eeden (1957a, 1958) and explain ways to use them to compute MLEs. We first need more notation. The MLE of θ when θ is restricted to $\Theta^* = \{\theta \mid \theta_i \in I_i, i = 1, \ldots, k\}$ will be denoted by $v = (v_1, \ldots, v_k)$. That MLE exists by Condition A and the MLE for θ restricted to Θ will be expressed in terms of these v_i.

First of all note that, when $\alpha_{i,j}(v_i - v_j) \leq 0$ for all $i < j$, then $v = (v_1, \ldots, v_k) \in \Theta$, implying that $\hat{\theta} = v$ in that case. So, in the sequel we suppose that there exists a pair (i, j) with $i < j$ and $\alpha_{i,j}(v_i - v_j) > 0$. Now suppose that $\{1, \ldots, k\} = M_1 \cup M_2$ with $M_1 \cap M_2 = \emptyset$ and, for $i_1 \in M_1$ and $i_2 \in M_2$, $\alpha_{i_1,i_2} = 0$ when $i_1 < i_2$ and $\alpha_{i_2,i_1} = 0$ when $i_2 < i_1$. Then the MLE of θ can be obtained by separately maximizing $\sum_{i \in M_1} l_i(\theta)$ and $\sum_{i \in M_2} l_i(\theta)$. So, in the sequel we suppose that such (M_1, M_2) do not exist.

Theorem 8.1 *If for some pair (i_1, i_2) with $i_1 < i_2$, we have $\alpha_{i_1,i_2}(v_{i_1} - v_{i_2}) > 0$ and*

$$\left. \begin{array}{ll} \alpha_{i_1,h} = \alpha_{h,i_2} = 0 & \text{for all } h \text{ between } i_1 \text{ and } i_2 \\[2mm] \alpha_{h,i_1} = \alpha_{h,i_2} & \text{for all } h < i_1 \\[2mm] \alpha_{i_1,h} = \alpha_{i_2,h} & \text{for all } h > i_2, \end{array} \right\} \tag{8.4}$$

then $\hat{\theta}_{i_1} = \hat{\theta}_{i_2}$.

This theorem says that, under its conditions, the problem of maximizing $l(\theta)$ for $\theta \in \Theta$ can be reduced to maximizing $l(\theta)$ for $\theta \in \Theta_1 = \{\theta \in \Theta \mid \theta_{i_1} = $

θ_{i_2}}. That for this new (k-1)-dimensional parameter space Condition A is satisfied follows from (8.4). Note that, for the simple order, (8.4) is satisfied for each (i_1, i_2) with $i_2 = i_1 + 1$. So the algorithm based on Theorem 8.1 is a generalization of the PAVA, which says that, for simply ordered θ_i, $\hat{\theta}_i = \hat{\theta}_{i+1}$ when $v_i > v_{i+1}$.

Theorem 8.2 *If, for a pair (i_1, i_2), $v_{i_1} \leq v_{i_2}$ and*

$$\left. \begin{array}{c} \alpha_{i_1,i_2} = 0 \\[2mm] \alpha_{h,i_1} \leq \alpha_{h,i_2} \text{ for all } h < i_1,\ h \neq i_2 \\[2mm] \alpha_{i_1,h} \geq \alpha_{i_2,h} \text{ or all } h > i_2,\ h \neq i_1, \end{array} \right\} \tag{8.5}$$

then $\hat{\theta}_{i_1} \leq \hat{\theta}_{i_2}$.

By this theorem one can, when its conditions are satisfied, add the restriction $\theta_{i_1} \leq \theta_{i_2}$, i.e., the problem can be reduced to maximizing $l(\theta)$ for $\theta \in \Theta_2 = \{\theta \in \Theta \mid \theta_{i_1} \leq \theta_{i_2}\}$. Obviously, Condition A is satisfied for Θ_2. An example where this theorem is very useful is the simple-tree order where $\alpha_{1,i} = 1$, $i = 2, \ldots, k$ and $\alpha_{i,j} = 0$ for $2 \leq i < j \leq k$. According to the theorem one can renumber the θ_i, $i = 2, \ldots, k$, in increasing order of their value of v_i and then solve the simple-order problem for those renumbered θ_i, for which the PAVA gives a simple solution. Thompson (1962) (see also Barlow, Bartholomew, Bremner and Brunk, 1972, p. 73–74 and Robertson, Wright and Dykstra, 1988, p. 57) gives the so-called minimum-violators algorithm for the rooted-tree order. For the simple-tree order, Thompson's (1962) algorithm reduces to the one of van Eeden (1957a, 1958). Another algorithm for the simple-tree-order case can be found in Qian (1996). He minimzes $\sum_{i=1}^{k} |g_i - \theta_i|$ for given $g_1, \ldots, g_k$, but his algorithm is less efficient than the one of van Eeden (1957a, 1958) based on Theorem 8.2.

Theorem 8.3 *Let, for a pair (i_1, i_2), $\alpha_{i_1,i_2} = 0$ and let $\Theta_3 = \{\theta \in \Theta \mid \theta_{i_1} \leq \theta_{i_2}\}$. Further, let $\hat{\theta}^*$ maximize $l(\theta)$ for $\theta \in \Theta_3$. Then $\hat{\theta}_i = \hat{\theta}^*$, $i = 1, \ldots, k$ when $\hat{\theta}^*_{i_1} < \hat{\theta}^*_{i_2}$ and $\hat{\theta}^*_{i_1} = \hat{\theta}^*_{i_2}$ when $\hat{\theta}^*_{i_1} \geq \hat{\theta}^*_{i_2}$.*

This theorem, like the foregoing one, changes the problem into one with more restrictions than Θ. It generalizes Theorem 8.2. But one needs to know $\hat{\theta}_i$ for at least $i = i_1$ and $i = i_2$ in order to be able to use it, whereas for Theorem 8.2 one only needs to know v_i for $i = i_1$ and $i = i_2$. On the other hand, Theorem 8.3 always applies no matter what the $\alpha_{i,j}$ are, while Theorem 8.2 only applies when the $\alpha_{i,j}$ satisfy (8.5)

There is a fourth theorem which, like Theorem 8.3, applies no matter what the $\alpha_{i,j}$ are. To state that theorem we need to define the so-called "essential restrictions" defining Θ. Those are restrictions $\theta_{i_1} \leq \theta_{i_2}$ satisfying $\alpha_{i_1,h}\alpha_{h,i_2} = 0$

for all h between i_1 and i_2. We denote them by $R_1, \ldots, R_s$. Each $\lambda \in \{1, \ldots, s\}$ corresponds to exactly one pair (i_1, i_2) which we denote by $(i_{\lambda,1}, i_{\lambda,2})$. Then $(R_1, \ldots, R_s)$ and $(\alpha_{i,j}(\theta_i - \theta_j) \leq 0, 1 \leq i \leq j \leq k)$ define the same subset of R^k.

Theorem 8.4 *If $\hat{\theta}' = (\hat{\theta}_1', \ldots, \hat{\theta}_k')$ maximizes l under the restrictions*

$$(R_1, \ldots, R_{\lambda-1}, R_{\lambda+1}, \ldots, R_s; \theta_i \in I_i, i = 1, \ldots, k),$$

then $\hat{\theta} = \hat{\theta}'$ when $\hat{\theta}'_{i_{\lambda,1}} \leq \hat{\theta}'_{i_{\lambda,2}}$. Further, $\hat{\theta}_{i_{\lambda,1}} = \hat{\theta}_{i_{\lambda,2}}$ when $\hat{\theta}'_{i_{\lambda,1}} > \hat{\theta}'_{i_{\lambda,2}}$.

With this theorem one reduces the number of restrictions by taking out one of the essential ones. If one can solve that problem, one either has found $\hat{\theta}$ or one has reduced the problem to that of maximizing l in a lower-dimensional space obtained from Θ by replacing an essential restriction by an equality. However, one needs to find at least $\hat{\theta}_{i_{\lambda,1}}$ and $\hat{\theta}_{i_{\lambda,2}}$ in order to be able to apply it.

As an example take the case where

$$\alpha_{1,2} = 1, \alpha_{1,4} = 1, \alpha_{2,3} = 1, \alpha_{3,4} = 0.$$

Suppose $v_1 \leq v_2$, $v_3 \leq v_4$ and $v_1 > v_4$. Then neither Theorem 8.1 nor Theorem 8.2 apply. But both Theorem 8.3 and Theorem 8.4 can be used to find the MLE. We first use Theorem 8.4 by taking out the essential restriction $\theta_1 \leq \theta_4$. This new problem is easily solved by using the PAVA on each of $\theta_1 \leq \theta_2$ and $\theta_3 \leq \theta_4$. If this gives a $\hat{\theta}'$ satisfying $\hat{\theta}_1' \leq \hat{\theta}_4'$, then $\hat{\theta} = \hat{\theta}'$. In case $\hat{\theta}_1' > \hat{\theta}_4'$, $\hat{\theta}$ maximizes l under the restrictions $\theta_3 \leq \theta_1 = \theta_4 \leq \theta_2$, a problem easily solved by the PAVA. Using Theorem 8.3 with $i_1 = 3$ and $i_2 = 4$, the PAVA can be used to find $\hat{\theta}^*$. This either solves the problem (namely, when $\hat{\theta}_3^* < \hat{\theta}_4^*$), or the problem with $\theta_1 \leq \theta_2 = \theta_3 \leq \theta_4$ needs to be solved. But that can be done by the PAVA.

Remark 8.1. Although each of the above theorems 8.2–8.4 can be found in van Eeden (1957a) as well as in van Eeden (1958), authors often refer to these van Eeden papers for what they call "van Eeden's algorithm" without specifying which one they mean. In fact they mean the one based on Theorem 8.4 and they comment on its efficiency. For example, Barlow, Bartholomew, Bremner and Brunk (1972, pp. 90–91) call it "rather complicated" and Dykstra (1981) calls it "inefficient" and suggests an improvement. Another improvement can be found in Gebhardt (1970). Further, Lee (1983) finds it "suitable for small problems" and finds it inefficient for large ones. I agree with Lee's (1983) statement and the example above shows it to be very handy for such small problems.

Brunk (1955) gives max-min formulas. For these, the following definitions are needed. A subset L of $\{1, \ldots, k\}$ is called a lower set if

$$(i \in L, \alpha_{j,i} = 1) \Rightarrow j \in L$$

and a subset U of $\{1, \ldots, k\}$ is called an upper set if

$$(i \in U, \alpha_{i,j} = 1) \Rightarrow j \in U.$$

Further, let $\mathcal{L}$ be the class of lower sets and $\mathcal{U}$ the class of upper sets. Then, for $i = 1, \ldots, k$, Brunk (1955) (see also Barlow, Bartholomew, Bremner and Brunk, 1972, p.80) shows that, under his conditions,

$$
\left.
\begin{aligned}
\hat{\theta}_i &= \max_{U \in \mathcal{U}} \min_{L \in \mathcal{L}} (v_{L \cap U} \mid i \in L \cap U) \\[2mm]
&= \min_{L \in \mathcal{L}} \max_{U \in \mathcal{U}} (v_{L \cap U} \mid i \in U \cap L) \\[2mm]
&= \max_{U \in \mathcal{U}} \min_{L \in \mathcal{L}} (v_{L \cap U} \mid i \in U, U \cap L \neq \emptyset) \\[2mm]
&= \min_{L \in \mathcal{L}} \max_{U \in \mathcal{U}} (v_{L \cap U} \mid i \in L, U \cap L \neq \emptyset).
\end{aligned}
\right\}
\tag{8.6}
$$

Robertson and Waltman (1968) prove (8.6) under their condition and show that, for the double–exponental case, one needs to define the median of a sample of even size as the average of the middle two of the ordered observations for their conditions to hold. Further, van Eeden (1957a, 1958) proves that the first one of the four formulas (8.6) holds under Condition A, but it can easily be seen that the other three of the formulas (8.6) also hold under Condition A.

Before presenting and comparing more algorithms for computing MLEs for the model and conditions described in Section 8.1, something needs to be said about the notion of "isotonic regression", a notion much used in order-restricted inference. It is discussed in Barlow, Bartholomew, Bremner and Brunk (1972). More can be found in Robertson, Wright and Dykstra (1988).

To introduce the notion here, suppose (in the model introduced in Section 8.1) that the $X_{i,j}$ are normally distributed with mean θ_i and variance 1. Then the MLE of $\theta = (\theta_1, \ldots, \theta_k)$ minimizes, for $\theta \in \Theta$,

$$l_N(\theta) = \sum_{i=1}^{k} \sum_{j=1}^{n_i} (X_{i,j} - \theta_i)^2 = \sum_{i=1}^{k} n_i (\bar{X}_i - \theta_i)^2 + \sum_{i=1}^{k} \sum_{j=1}^{n_i} (X_{i,j} - \bar{X}_i)^2,$$

where $n_i \bar{X}_i = \sum_{i=1}^{n_i} X_{i,j}$, $i = 1, \ldots, k$. So, in this case, the MLE of θ for $\theta \in \Theta$ minimizes the weighted squared distance between $\bar{X} = (\bar{X}_1, \ldots, \bar{X}_k)$ and Θ. This MLE is called the (weighted) isotonic (least-squares) regression of $\bar{X}$ with respect to the ordering of the θ_i implied by $\theta \in \Theta$.

Now look at the case where the $X_{i,j}$ have the double-exponential density $e^{-|x - \theta_i|}/2$, $-\infty < x < \infty$. Then the MLE of θ for $\theta \in \Theta$ minimizes

$l_{DE}(\theta) = \sum_{i=1}^{k} \sum_{j=1}^{n_i} |X_{i,j} - \theta_i|$. Some authors call this MLE the "isotonic median regression" of the $X_{i,j}$ with respect to the ordering of the θ_i implied by $\theta \in \Theta$ (see, e.g., Menéndez and Salvador, 1987, and Chakravarti, 1989). Note, however, that minimizing l_{DE} is not equivalent to minimizing $\sum_{i=1}^{k} |M_i - \theta_i|$, nor to minimizing $\sum_{i=1}^{k} n_i |M_i - \theta_i|$, where M_i is the median of $X_{i,j}$, $j = 1, \ldots, n_i$. So, here the MLE is not the closest $\theta \in \Theta$ (in the least-absolute-deviation (LAD) sense) to $(M_1, \ldots, M_k)$. Some authors call the regression minimizing $\sum_{i=1}^{k} |M_i - \theta_i|$ the "isotonic LAD regression" (see, e.g., Qian (1994a), who claims this estimator to be the MLE in case of double-exponential observations); others call it the "isotonic median regression" (see, e.g., Cryer, Robertson, Wright and Casady, 1972). I will call the θ minimizing $\sum_{i=1}^{k} \sum_{j=1}^{n_i} |X_{i,j} - \theta_i|$ the isotonic median regression and the one minimizing $\sum_{i=1}^{k} |M_i - \theta_i|$ the isotonic LAD regression.

The isotonic LAD regression has been generalized to the "isotonic percentile regression" by Casady and Cryer (1976). It minimizes $\sum_{i=1}^{k} |\hat{\alpha}_i - \theta_i|$, where, for $i = 1, \ldots, k$, and a given $\alpha \in (0,1)$, $\hat{\alpha}_i$ is the 100α-th percentile of $X_{i,1}, \ldots, X_{i,n_i}$.

Note that, for defining estimators of $\theta \in \Theta$, the function $l(\theta)$ does not have to be the likelihood function of the $X_{i,j}$. One can maximize any function $l^*(\theta)$ defined on $\prod_{i=1}^{k} J_i$ and satisfying Condition A. However, these estimators are not necessarily the MLE of θ for $\theta \in \Theta$, a fact that, particularly when l is a weighted sum of squares, does not seem to be generally known. An example is the case where $X_{i,j} \sim^{ind} \mathcal{U}(0, \theta_i)$, $j = 1, \ldots, n_i$, $i = 1, \ldots, k$. For this case Gupta and Leu (1986) claim that the isotonic least-squares regression gives the MLE, which is clearly not true. Further, Jewel and Kalbfleisch (2004, p. 303) state, in part, that "In one dimension maximum likelihood is equivalent to least-squares ...". But maybe these authors mean that this equivalence holds for the one-dimensional version of their problem, i.e., the estimation of a restricted θ based on $X_i \sim^{ind} \text{Bin}(n_i, \theta_i)$, $i = 1, \ldots, k$. For that case, maximum likelihood is equivalent to weighted least-squares with weights $n_1, \ldots, n_k$. I will, however, keep using the notation $\hat{\theta}$ for the maximizer of a function satisfying Condition A, even when this maximizer is not the MLE.

A recursion formula for isotonic least-squares regression can be found in Puri and Singh (1990). They show that the minimizer of $\sum_{i=1}^{k} (g_i - \theta_i)^2 w_i$ for given numbers g_i, $i = 1, \ldots, k$, and given positive numbers w_i, $i = 1, \ldots, k$, is given by

$$\hat{\theta}_1 = \min_{1 \le i \le k} \frac{G_i}{W_i}$$

$$\hat{\theta}_j = \min_{j \le i \le k} \frac{G_i - \sum_{r=1}^{j-1} w_r \hat{\theta}_r}{\sum_{r=j}^{i} w_r}, \qquad j = 2, \ldots, k,$$

where, for $j = 1, \ldots, k$, $G_j = \sum_{i=1}^{j} g_i w_i$ and $W_j = \sum_{i=1}^{j} w_i$. They prove this result by relying on the so-called greatest convex minorant algorithm (see Barlow, Bartholomew, Bremner and Brunk, 1972, pp. 9–13). Their proof is correct, but they do not seem to know that the greatest convex minorant algorithm only holds for simply ordered θ_i.

An algorithm not yet mentioned is the so-called "minimum lower sets algorithm". For least-squares isotonic regression it is given by Brunk (1955) and Brunk, Ewing and Utz (1957) (see also Barlow, Bartholomew, Bremner and Brunk, 1972, pp. 76–77, and Robertson, Wright and Dykstra, 1988, pp. 24–25). But it holds more generally. In fact, it gives the maximizer of $l(\theta)$ when l satisfies Condition A and has the strict CMV property. It works as follows. Find a lower set L satisfying $v_L \le v_{L'}$ for all lower sets L'. If there is more than one such lower set, take their union. Call this union L_1. Then $\hat{\theta}_i = v_{L_1}$ for $i \in L_1$. Now repeat this process of finding such minimum lower sets by next looking at $\{i \in \{1, \ldots, k\} \mid i \in L_1^c\}$ and continue until all $\hat{\theta}_i$ have been found. There is, of course, also a maximum upper set algorithm.

Robertson and Wright (1973) use this minimum lower sets algorithm for isotonic median regression, where l does not have the strict CMV property. But they correct this mistake in Robertson and Wright (1980) where they give a minimum lower sets algorithm for the case where l has the CMV property, but not necessarily the strict CMV property.

Neither one of these minimum lower sets algorithms is very efficient, because finding the largest minimum lower set can be quite cumbersome. Qian (1992) improves on each of the above minimum lower set algorithms. He finds that the largest minimum lower set can be replaced by any minimum lower set.

In Strömberg (1991) an algorithm is given for finding all isotonic monotone regressions obtained by minimizing $\sum_{i=1}^{k} \sum_{j=1}^{n_i} d(X_{i,j} - \theta_i)$ under the restriction $\theta_1 \le \ldots, \le \theta_k$. He supposes the distance function d to be convex and shows that if the function d is strictly convex, the minimizer is unique. Note here that, in case of strict convexity, $l^*(\theta_i) = -\sum_{j=1}^{n_i} d(X_{i,j} - \theta_i)$ satisfies Condition A and the unicity of the minimizer then follows from the results of van Eeden (1957a, 1958).

Park (1998) gives, for given positive numbers $w_{i,j}$ and given $x_{i,j}$, the minimizer (in $\theta_{i,j}$) of

$$\sum_{j=1}^{k} \sum_{i=1}^{2} w_{i,j} (x_{i,j} - \theta_{i,j})^2$$

under the restriction that the $\Delta_j = \theta_{1,j} - \theta_{2,j}$, $j = 1, \ldots, k$ are simply ordered. He finds, for the minimizer $(\hat{\Delta}_1, \ldots, \hat{\Delta}_k, \hat{\theta}_{2,1}, \ldots, \hat{\theta}_{2,k})$, that $(\hat{\Delta}_1, \ldots, \hat{\Delta}_k)$ is the weighted (with weights $c_{i,j} = w_{1,j} w_{2,j} / (w_{1,j} + w_{2,j})$) isotonic least-squares regression of the $x_{1,j} - x_{2,j}$ with respect to the ordering of the Δ_j, while $\hat{\theta}_{2,j} = (w_{1,j}(x_{1,j} - \hat{\Delta}_j) + w_{2,j} x_{2,j}) / (w_{1,j} + w_{2,j})$, $j = 1, \ldots, k$. He applies his result to the problem of maximum likelihood estimation of the $\mu_{i,j}$ and the $\Delta_j = \mu_{1,j} - \mu_{2,j}$ in the model

$$Y_{i,j,l} = \mu_{i,j} + \varepsilon_{i,j,l}, \quad l = 1, \ldots, n_{i,j}, j = 1, \ldots, k, i = 1, 2,$$

where $\varepsilon_{i,j,l} \sim^{ind} \mathcal{N}(0, \sigma^2)$, $\mu_{i,j} = \mu + \alpha_i + \beta_j + \gamma_{i,j}$ with $\sum_{i=1}^2 \alpha_i = \sum_{j=1}^k \beta_j = \sum_{i=1}^2 \gamma_{i,j} = \sum_{j=1}^k \gamma_{i,j} = 0$ and $\gamma_{1,1} \leq \ldots \leq \gamma_{1,k}$. His algorithm applies here because the restriction on the $\gamma_{i,j}$ is equivalent to $\Delta_1 \leq \ldots \leq \Delta_k$ and isotonic least-squares regression with $w_{i,j} = n_{i,j}$ is equivalent to maximum likelihood because the $Y_{i,j,l}$ are independent $\mathcal{N}(\mu_{i,j}, \sigma^2)$, $l = 1, \ldots, n_{i,j}$.

One last remark on algorithms. Both Chakravarti (1989) and X. Hu (1997) present algorithms for cases where $I_i \neq J_i$ for some $i \in \{1, \ldots, k\}$. Chakravarti (1989, p. 136) does not seem to be aware of the fact that van Eeden's (1957a, 1958) algorithms for such cases apply to more than weighted squared-error loss. The algorithm of X. Hu (1997) is different from van Eeden's. He first finds the maximizer $\hat{\theta}^*$ of l for θ in the larger space $\Theta^* = \{\theta \mid \alpha_{i,j}(\theta_i - \theta_j) \leq 0, 1 \leq i < j \leq k\}$ and then finds $\hat{\theta}$ from $\hat{\theta}^*$. The van Eeden algorithm first maximizes l for $\theta \in \prod_{i=1}^k I_i$ and then obtains $\hat{\theta}$ from that maximizer – which seems to me to be the more efficient way of doing things.

8.3 Norm-reducing properties of MLEs

For the problem of estimating simply ordered probabilities, i.e., for the case where the $X_{i,j} \sim \text{Bin}(1, \theta_i)$, $i = 1, \ldots, k$, Ayer, Brunk, Ewing, Reid and Silverman (1955) note that, for all $\theta \in \Theta$,

$$\sum_{i=1}^k \left(\frac{X_i}{n_i} - \theta_i \right)^2 n_i \geq \sum_{i=1}^k (\hat{\theta}_i - \theta_i)^2 n_i + \sum_{i=1}^k \left(\frac{X_i}{n_i} - \hat{\theta}_i \right)^2 n_i,$$

where $X_i = \sum_{j=1}^{n_i} X_{i,j}$, $i = 1, \ldots, k$ and $\hat{\theta} = (\hat{\theta}_1, \ldots, \hat{\theta}_k)$ is the MLE of θ. This shows that this MLE minimizes $\sum_{i=1}^k ((X_i/n_i) - \theta_i)^2 n_i$ and that $\hat{\theta}$ is, in this weighted least-squares sense, closer to Θ than is the unrestricted MLE. Or, to say it another way, the MLE of θ is the weighted isotonic (least-squares) regression of $(X_1/n_1, \ldots, X_k/n_k)$ with respect to the simple order of the θ_i's. These authors implicitly assume that $I_i = J_i$ for all $i \in \{1, \ldots, k\}$.

A more general result can be found in van Eeden (1957c). She considers the model as described in Section 8.1. She does not need $I_i = J_i$, $i = 1, \ldots, k$ and assumes that Condition A is satisfied. She gives sufficient conditions for the MLE to be the minimizer of

$$Q(\theta_1, \ldots, \theta_k) = \sum_{i=1}^{k} w_i (\theta_i - v_i^*)^2,$$

where, for $i = 1, \ldots, k$, v_i^* is the MLE of θ_i under the condition that $\theta_i \in J_i$. Further, the w_i are positive numbers satisfying the following condition. Let E_o be the set of all $i \in \{1, \ldots, k\}$ for which $\hat{\theta}_i \neq v_i$ and let M_o be a subset of E_o for which

$$\hat{\theta}_i = \hat{\theta}_j \text{ for all } i, j \in M_o. \tag{8.7}$$

Then the condition on the w_i is that

$$\sum_{i \in M_o} w_i (v_{M_o}^* - v_i^*) = 0 \quad \text{for all } M_o \subset \{1, \ldots, k\} \text{ satifying (8.7).} \tag{8.8}$$

Of course, these w_i need to be independent of the v_i^*. In van Eeden (1957c) several examples are given where w_i satisfying the above conditions exist and can be explicitly obtained. The examples she mentions are

a) $X_{i,j} \sim \mathcal{N}(\theta_i, \sigma_i^2)$ with known variances σ_i^2, where $v_i^* = \bar{X}_i$ and $w_i = n_i/\sigma_i^2$;
b) $X_{i,j} \sim Bin(1, \theta_i)$, where $v_i^* = X_i/n_i$ and $w_i = n_i$;
c) $X_{i,j} \sim \mathcal{N}(0, \theta_i)$, where $v_i^* = \sum_{j=1}^{n_i} X_{i,j}^2/n_i$ and $w_i = n_i$;
d) $X_{i,j}$, $j = 1, \ldots, n_i$ have an exponential distribution with density $e^{-x/\theta_i}/\theta_i$ on $(0, \infty)$ where $v_i^* = \sum_{i=1}^{n_i} X_{i,j}/n_i$ and $w_i = n_i$.

Note that a sufficient condition for (8.8) to hold is that $\sum_{i \in M} w_i (v_M^* - v_i^*) = 0$ for all $M \subset \{1, \ldots, k\}$. This last condition is the one given by Qian (1994b) in his Theorem 2.1, but he seems to suppose that $I_i = J_i$ for all $i = 1, \ldots, k$.

Robertson, Wright and Dykstra (1988, Theorem 1.5.2) give an explicit condition for the MLE to be the isotonic least-squares regression of the v_i^*. They suppose that $X_{i,j}$, $j = 1, \ldots, n_i$, $i = 1, \ldots, k$ are independent samples from exponential-family distributions with densities (with respect to a σ-finite measure ν) $f(x; \theta_i, \tau_i)$, where

$$f(x; \theta, \tau) = exp\{p_1(\theta)p_2(\tau)K(x; \tau) + S(x; \tau) + q(\theta, \tau)\}.$$

Here, τ is a nuissance parameter taking values in T and $\theta \in J$, the natural parameter space of the family, is the parameter of interest. Under regularity conditions on p_1, p_2 and q and assuming that

$$\frac{d}{d\theta} q(\theta, \tau) = -\theta \frac{d}{d\theta} p_1(\theta) \, p_2(\tau) \quad \text{for all } \theta \in N, \tau \in T,$$

they show that, when $I_i = J$ for all $i = 1, \ldots, k$, the MLE $\hat{\theta}$ of θ minimizes, for $\theta \in \Theta$,

$$\sum_{i=1}^{k} n_i p_2(\tau)(v_i^* - \theta_i)^2.$$

Note that, for each of the examples a) – d) above, these conditions are satisfied with $p_2(\tau_i) = \sigma_i^{-2}$ for the $\mathcal{N}(\theta_i, \sigma_i)$ distribution and with $p_2(\tau_i) = 1$ in the other cases. But van Eeden (1957c) does not need $I_i = J_i$ for all $i = 1, \ldots, k$.

Further, X. Hu (1997) shows that the Robertson–Wright–Dykstra result also holds when when $I_i = I \subset J$, $I \neq J$, $i = 1, \ldots, k$.

And, finally, note that for $X_{i,j} \sim \mathcal{N}(0, \theta_i^2)$, neither the conditions of Robertson, Wright and Dykstra (1988, Theorem 1.5.2) nor the condition (8.8) of van Eeden (1957c) is satisfied. These conditions are also not satisfied for $X_{i,j} \sim \mathcal{U}(0, \theta_i)$. However, a "random" version of (8.8) (more precisely of its sufficient condition $\sum_{i \in M} w_i(v_M^* - v_i^*) = 0$ for all $M \subset \{1, \ldots, k\}$) is satisfied for this uniform case. Take, e.g., $k = 2$, $I_i = J_i$, $i = 1, 2$ and let $\Theta = \{\theta \mid \theta_1 \leq \theta_2\}$. Then we only need to look at the case where $M = \{1, 2\}$ and in that case the likelihood function needs to be maximized over the random set $\{\theta \mid v_1^* \leq \theta_1 \leq \theta_2, v_2^* \leq \theta_2\}$ when $v_1^* \leq v_2^*$ and over the random set $\{\theta \mid v_1^* \leq \theta_1 \leq \theta_2\}$ when $v_2^* < v_1^*$. Clearly, this is equivalent to minimizing $\sum_{i=1}^{2}(\theta_i - v_i^*)^2$ over these sets.

8.4 Algorithms for multivariate problems

Sasabuchi, Inutsuka and Kulatunga (1983) consider the case where, for $i = 1, \ldots, k$, $\theta_i = (\theta_{1,i}, \ldots, \theta_{p,i})'$ for some $p \geq 2$. Their parameter space is defined by

$$\Theta = \{(\theta_1, \ldots, \theta_k) \mid \alpha_{i,j}(\theta_{\nu,i} - \theta_{\nu,j}) \leq 0, \nu = 1, \ldots, p, 1 \leq i < j \leq k\}$$

where the $\alpha_{i,j}$ are independent of ν and satisfy the conditions of Section 8.1. Then, for positive definite $p \times p$ matrices $\Lambda_1, \ldots, \Lambda_k$ and a $p \times k$ matrix $X = (X_1, \ldots, X_k)$, they define $\hat{\theta} = (\hat{\theta}_1, \ldots, \hat{\theta}_k)$ to be the p-variate isotonic regression of X with respect to the weights $\Lambda_1^{-1}, \ldots, \Lambda_k^{-1}$ if $\hat{\theta}$ minimizes, for $\theta \in \Theta$,

$$\sum_{i=1}^{k}(X_i - \theta_i)' \Lambda_i^{-1}(X_i - \theta_i).$$

They give an algorithm for computing $\hat{\theta}$ for the case when $p = 2$. It consists of iterative applications of univariate isotonic regressions. They also indicate how this algorithm can be extended to the case $p > 2$. Of course, when all the Λ_i are the identity matrix, $\hat{\theta}$ can be obtained by finding, for each $\nu = 1, \ldots, p$, the isotonic regression of $(X_{\nu,1}, \ldots, X_{\nu,k})$ with respect to equal weights. As an example they look at the case where $X_i \sim^{ind} \mathcal{N}_p(\theta_i, \Lambda_i)$, $i = 1, \ldots, k$, $p \geq 2$. In this case $\hat{\theta}$ is, of course, the MLE of θ based on X under the restriction $\theta \in \Theta$.

Another multivariate case can be found in Jewel and Kalbfleisch (2004). They consider the case where $X_1, \ldots, X_k$ are independent and, for $i = 1, \ldots, k$, X_i has a multinomial distribution with parameters n_i and $\theta_i = (\theta_{1,i}, \ldots, \theta_{p,i})$, where, for $i = 1, \ldots, k$, $\sum_{\nu=1}^{p} \theta_{\nu,i} = 1$. The parameter space is restricted by the inequalities $\theta_{\nu,1} \leq \ldots, \leq \theta_{\nu,k}$ for each $\nu \in \{1, \ldots, p-1\}$ and they give an algorithm for finding the MLE. For the case where $p = 2$, i.e., when $X_i \sim^{ind}$ $\mathrm{Bin}(n_i, \theta_i)$, $i = 1, \ldots, k$ with $\theta_1 \leq \ldots \leq \theta_k$, their algorithm reduces to the PAVA.

Bibliography[1]

ABELSON, R.P. and TUKEY, J.W. (1963). Efficient utilization of non-numerical information in quantitative analysis: General theory and the case of simple order. Ann. Math. Statist., 34, 1347–1369.

AKKERBOOM, J.C. (1990). *Testing Problems with Linear and Angular Inequality Constraints*. Lecture Notes in Statistics, Vol. 62. Springer-Verlag.

ALI, M.M. and WOO, J. (1998). Bayes estimation of Bernoulli parameter in restricted parameter space. J. Statist. Res., 32, 81–87.

AL-SALEH, M.F. (1997). Estimating the mean of a normal population utilizing some available information: A bayesian approach. J. Information & Optimization Sciences, 18, 1–7.

*ALSON, P. (1988). Minimax properties for linear estimators of the location parameter of a linear model. Statistics, 19, 163–171. (Gaffke, Heiligers (1991) show that Theorem 2.2 in this paper is incorrect.)

*ALSON, P. (1993a). Linear minimaxity and admissibility for centered bounded or unbounded ellipsoids. Rebrape, 7, 201–217.

*ALSON, P. (1993b). Centered ellipsoids for which an admissible linear estimator is the minimax linear estimator. Statistics, 24, 85–94.

AYER, M., BRUNK, H.D., EWING, G.M., REID, W.T. and SILVERMAN, E. (1955). An empirical distribution function for sampling with incomplete information. Ann. Math. Statist., 26, 641–647.

BADER, G. and BISCHOFF, W. (2003). Old and new aspects of minimax estimation of a bounded parameter. In *Mathematical Statistics and Applications: Festschrift for Constance van Eeden* (M. Moore, S. Froda and C. Léger, eds.), IMS Lecture Notes and Monograph Series, 43, 15–30. Institute of Mathematical Statistics, Hayward, California, USA.

BARLOW, R.E., BARTHOLOMEW, D.J., BREMNER, J.M. and BRUNK, H.D. (1972). *Statistical Inference under Order Restrictions. The Theory and Application of Isotonic Regression*. John Wiley & Sons.

---

[1] Papers marked with an asterisk treat (see Chapter 2) the $(\mathcal{D}_o, \Theta)$-problem, mostly for the linear model and with linear estimators. See Chapter 6 for details.

BERGER, J. (1984). The robust Bayesian viewpoint. In *Robustness of Baysian Analyses* (J Kadane, ed.), 63–124. Elsevier Science Publishers.

BERGER, J.O. (1985). *Statistical Decision Theory and Bayesian Analysis.* Second Edition. Springer-Verlag.

BERRY, J.C. (1989). Bayes minimax estimation of a Bernoulli p in a restricted parameter space. Comm. Statist. Theory Methods, 18, 4607–4616.

BERRY, J.C. (1990). Minimax estimation of a bounded normal mean vector. J. Multivariate Anal., 35, 130–139.

BERRY, J.C. (1993). Minimax estimation of a restricted exponential location parameter. Statist. Decisions, 11, 307–316.

BERRY, J.C. (1994). Improving the James-Stein estimator using the Stein variance estimator. Statist. Probab. Lett., 20, 241–245.

BISCHOFF, W. (1992). Minimax and Γ-minimax estimation for functions of the bounded parameter of a scale parameter family under "L_p-loss". Statist. Decisions, 10, 45–61.

BISCHOFF, W. and FIEGER, W. (1992). Minimax estimators and Γ-minimax estimators for a bounded normal mean under the loss $l_p(\theta, d) = |\theta - d|^p$. Metrika, 39, 185–197.

BISCHOFF, W. and FIEGER, W. (1993). On least favourable two point priors and minimax estimators under absolute error loss. Metrika, 40, 283–298.

BISCHOFF, W., FIEGER, W. and OCHTROP, S. (1995). Minimax estimation for the bounded mean of a bivariate normal distribution. Metrika, 42, 379–394.

BISCHOFF, W., FIEGER, W. and WULFERT, S. (1995). Minimax- and Γ-minimax estimation of a bounded normal mean under linex loss. Statist. Decisions, 13, 287–298.

BLUMENTHAL, S. and COHEN, A. (1968a). Estimation of the larger translation parameter. Ann. Math. Statist., 39, 502–516.

BLUMENTHAL, S. and COHEN, A. (1968b). Estimation of two ordered translation parameters. Ann. Math. Statist., 39, 517–530.

BLYTH, C.R. (1951). On minimax statistical decision procedures and their admissibility. Ann. Math. Statist., 22, 22–42.

BLYTH, C.R. (1993). Restrict estimates to the possible values ? Amer. Statist., 47, 73–75.

BOROVKOV, A. and SAKHANIENKO, A. (1980). On estimates of the expected quadratic risk. Probab. Math. Statist., 1, 185–195 (in Russian).

BREWSTER, J.F. and ZIDEK, J.V. (1974). Improving on equivariant estimators. Ann. Statist., 2, 21–38.

BROWN, L.D. (1981). A complete class theorem for statistical problems with finite sample spaces. Ann. Statist., 9, 1289–1300.

BROWN, L.D. (1986). *Fundamentals of Statistical Exponential Families with Applications in Statistical Decision Theory.* Lecture Notes-Monograph Series, Vol. 9., Institute of Mathematical Statistics, Hayward, California, USA.

BROWN, L.D., CHOW, M and FONG, D.K.H. (1992). On the admissibility of the maximum-likelihood estimator of the binomial variance. Canad. J. Statist., 20, 353–358.

BROWN, L.D. and GAJEK, L. (1990). Information inequalities for the Bayes risk. Ann. Statist., 18, 1578–1594.

BROWN, L.D. and LOW, M.G. (1991). Information inequality bounds on the minimax risk (with an application to nonparametric regression). Ann. Statist., 19, 329–337.

BRUNK, H.D. (1955). Maximum likelihood estimates of monotone parameters. Ann. Math. Statist., 26, 607–616.

BRUNK, H.D. (1958). On the estimation of parameters restricted by inequalities. Ann. Math. Statist., 29, 437–454.

BRUNK, H.D., EWING, G.M. and UTZ, W.R. (1957). Minimizing integrals in certain classes of monotone functions. Pacific J. Math., 7, 833–847.

*BUNKE, O. (1975a). Minimax linear, ridge and shrunken estimators for linear parameters. Math. Operationsforsch. u. Statist., 6, 697–701.

*BUNKE, O. (1975b). Improved inference in linear models with additional information. Math. Operationsforsch. u. Statist., 6, 817–829.

CASADY, R.J. and CRYER, J.D. (1976). Monotone percentile regression. Ann. Statist., 4, 532–541.

CASELLA, G. and STRAWDERMAN, W.E. (1981). Estimating a bounded normal mean. Ann. Statist., 9, 870–878.

CHAKRAVARTI, N. (1989). Bounded isotonic median regression. Comput. Statist. Data Anal., 8, 135–142.

CHANG, Y.-T. (1981). Stein-type estimators for parameters restricted by linear inequalities. Keio Science and Technology Reports, 34, 83–95.

CHANG, Y.-T. (1982). Stein-type estimators for parameters in truncated spaces. Keio Science and Technology Reports, 35, 185–193.

CHANG, Y.-T. (1991). Stein-type estimators for two parameters under order restriction. J. Tokiwa Univ., Vol. 20, 87–92.

CHANG, Y.-T. and SHINOZAKI, N. (2002). A comparison of restricted and unrestricted estimators in estimating linear functions of ordered scale parameters of two gamma distributions. Ann. Inst. Statist. Math., 54, 848–860.

CHARRAS, A. (1979). *Propriété Bayesienne et Admissibilité d'Estimateurs dans un Sous-ensemble Convex de R^p*. PhD. thesis, Université de Montréal, Montréal, Canada.

CHARRAS, A. and van EEDEN, C. (1991a). Bayes and admissibility properties of estimators in truncated parameter spaces. Canad. J. Statist., 19, 121–134.

CHARRAS, A. and van EEDEN, C. (1991b). Limits of Bayes estimators in convex, truncated parameter spaces. Statist. Probab. Lett., 11, 479–483.

CHARRAS, A. and van EEDEN, C. (1992). Bayes properties of estimators of location parameters in truncated parameter spaces, Statist. Decisions, 10, 81–86.

CHARRAS, A. and VAN EEDEN, C. (1994). Inadmissibility for squared error loss when the parameter to be estimated is restricted to the interval $[a, \infty)$. Statist. Decisions, 12, 257–266.

*CHATURVEDI, A. and WAN, A.T.K. (1999). Estimation of regression coefficients subject to interval constraints in models with non-spherical errors. Sankhyā Ser. B, 61, 433–442.

CHEN, L. and EICHENAUER, J. (1988). Two point priors and Γ-minimax estimating in families of uniform distributions. Statist. Papers, 29, 45–57.

*CHRISTOPEIT, N. and HELMES, K. (1996). Linear minimax estimation with ellipsoidal constraints. Acta Appl. Math., 43, 3–15.

COHEN, A. and KUSHARY, D. (1998). Universal admissibility of maximum likelihood estimators in constrained spaces. Statist. Decisions, 16, 131–146.

COHEN, A. and SACKROWITZ, H.B. (1970). Estimation of the last mean of a monotone sequence. Ann. Math. Statist., 41, 2021–2034.

CRYER, J.D., ROBERTSON, T., WRIGHT, F.T. and CASADY, R.J. (1972). Monotone median regression. Ann. Math. Statist., 43, 1459–1469.

DASGUPTA, A. (1985). Bayes minimax estimation in multiparameter families when the parameter space is restricted to a bounded convex set. Sankhyā Ser. A, 47, 326–332.

DONOHO, D.L., LIU, R.C., MACGIBBON, B. (1990). Minimax risks over hyperrectangles and implications. Ann. Statist., 18, 1416–1437.

*DRYGAS, H. (1996). Spectral methods in linear minimax estimation. Acta Appl. Math., 43, 17–42.

*DRYGAS, H. (1999). Linear minimax estimation in the three parameters case. Tatra Mt. Math. Publ., 17, 311–318.

DYKSTRA, R.L. (1981). An isotonic regression algorithm. J. Statist. Plann. Inference, 5, 355–363.

DYKSTRA, R. (1990). Minimax estimation of a mean vector for distributions on a compact set. Astin Bull., 20, 173–179.

DZO-I, D. (1961). O minimaksowej estymacji parametru rozkladu dwumianowego. (Translation: On the minimax estimation of the parameter of a binomial distribution.), Zastos Mat., 6, 31–42.

EICHENAUER, J. (1986). Least favourable two point priors in estimating the bounded location parameter of a noncentral exponential distribution. Statist. Decisions, 4, 389–392.

EICHENAUER, J., KIRSCHGARTH, P. and LEHN, J. (1988). Gamma-minimax estimation for a bounded normal mean. Statist. Decisions, 6, 343–348.

EICHENAUER-HERRMANN, J. and FIEGER, W. (1989). Minimax estimation in scale parameter families when the parameter interval is bounded. Statist. Decisions, 7, 363–376.

EICHENAUER-HERRMANN, J. and FIEGER, W. (1992). Minimax estimation under convex loss when the parameter interval is bounded. Metrika, 39, 27–43.

EICHENAUER-HERRMANN, J. and ICKSTADT, K. (1992). Minimax estimators for a bounded location parameter. Metrika, 39, 227–237.

ELFESSI, A. and PAL, N. (1992). A note on the common mean of two normal populations with order restricted variances. Comm. Statist. Theory Methods, 21, 3177–3184.

ESCOBAR, L.A. and SKARPNESS, B. (1987). Mean square error and efficiency of the least squares estimator over interval constraints. Comm. Statist. Theory Methods, 16, 397–406.

FAREBROTHER, R.W. (1975). The minimum mean square error linear estimator and ridge regression. Technometrics, 17, 127–128.

FARRELL, R.H. (1964). Estimators of a location parameter in the absolutely continuous case. Ann. Math. Statist., 35, 949–998.

FERNÁNDEZ, M.A., RUEDA, C. and SALVADOR, B. (1997). On the maximum likelihood estimator under order restrictions in uniform probability models. Comm. Statist. Theory Methods, 26, 1971–1980.

FERNÁNDEZ, M.A., RUEDA, C. and SALVADOR, B. (1998). Simultaneou estimation by isotonic regression. J. Statist. Plann. Inference, 70, 111–119.

FERNÁNDEZ, M.A., RUEDA, C. and SALVADOR, B. (1999). The loss of efficiency estimating linear functions under restrictions. Scand. J. Statist., 26, 579–592.

FERNÁNDEZ, M.A., RUEDA, C. and SALVADOR, B. (2000). Parameter estimation under orthant restrictions. Canad. J. Statist., 28, 171–181.

FOURDRINIER, D., OUASSOU, I. and STRAWDERMAN, W.E. (2003). Estimation of a parameter vector when some of the components are restricted. J. Multivariate Anal., 86, 2003, 14–27.

FUNO, E. (1991). Inadmissibility results of the MLE for the multinomial problem when the parameter space is restricted or truncated. Comm. Statist. Theory Methods, 20, 2863–2880.

*GAFFKE, N. and HEILIGERS, B. (1989). Bayes, admissible, and minimax linear estimators in linear models with restricted parameter space. Statistics, 20, 487–508.

*GAFFKE, N. and HEILIGERS, B. (1991). Note on a paper by P. Alson. Statistics, 22, 3–8.

GAJEK, L. (1987). An improper Cramer-Rao bound. Zastos, XIX, 241–256.

GAJEK, L. an KAŁUSZKA, M. (1995). Nonexponential applications of a global Cramér-Rao inequality. Statistics, 26, 111–122.

GARREN, S.T. (2000). On the improved estimation of location parameters subject to order restrictions in location-scale families. Sankhyā Ser. B, 62, 189–201.

GATSONIS, C., MacGIBBON, B. and STRAWDERMAN, W. (1987). On the estimation of a restricted normal mean. Statist. Probab. Lett., 6, 21–30.

GEBHARDT, F. (1970). An algorithm for monotone regression with one or more independent variables. Biometrika, 57, 263–271.

GHOSH, J.K. and SINGH, R. (1970). Estimation of the reciprocal of scale parameter of a gamma density. Ann. Inst. Statist. Math., 22, 51–55.

GHOSH, K. and SARKAR, S.K. (1994). Improved estimation of the smallest variance. Statist. Decisions, 12, 245–256.

GILL, R.D. and LEVIT, B.Y. (1995). Applications of the van Trees inequality: a Bayesian Cramér-Rao bound. Bernoulli, 1, 59–79.

*GIRKO, V.L. (1996). Spectral theory of minimax estimation. Acta Appl. Math., 43, 59–69.

GOURDIN, E., JAUMARD, B. and MACGIBBON, B. (1990). *Global optimization decomposition methods for bounded parameter minimax estimation.* Tech. Report G-90-48, GERARD, École des Hautes Études Commerciales, Montréal, Canada.

GRAYBILL, F.A. and DEAL, R.B. (1959). Combining unbiased estimators. Biometrics, 15, 543–550.

GREENBERG, B.G., ABUL-ELA, A-L.A., SIMMONS, W.R. and HORVITZ, D.G. (1969). The unrelated question randomized response model: Theoretical framework. J. Amer. Statist. Assoc., 64, 520–539.

*GROSS, J. (1996). Estimation using the linear regression model with incomplete ellipsoidal restrictions. Acta Appl. Math., 43, 81–85.

GUPTA, R.D. and SINGH, H. (1992). Pitman nearness comparisons of estimates of two ordered normal means. Austral. J. Statist., 34, 407–414.

GUPTA, S.S. and LEU, L.-Y. (1986). Isotonic procedures for selecting populations better than a standard for two-parameter exponential distributions. In *Reliability and Quality Control* (A.P. Basu, ed.), 167–183. Elsevier Science Publishers (North Holland).

HARTIGAN, J.A. (2004). Uniform priors on convex sets improve risk. Statist. Probab. Lett., 67, 285–288.

HASEGAWA, H. (1991). The MSE of a pre-test estimator of the linear regression model with interval constraints on coefficients. J. Japan Statist. Soc., 21, 189–195.

*HEILIGERS, B. (1993). Linear Bayes and minimax estimation in linear models with partially restricted parameter space. J. Statist. Plann. Inference, 36, 175–184.

HENGARTNER, N.W. (1999). A note on maximum likelihood estimation. Amer. Statist., 53, 123–125.

*HERING, F., TRENKLER, G. and STAHLECKER, P. (1987). Partial minimax estimation in regression analysis. Statist. Neerlandica, 41, 111–128.

HODGES, J.L.Jr. and LEHMANN, E.L. (1950). Some problems in minimax point estimation. Ann. Math. Statist., 21, 182–197.

HODGES, J.L.Jr. and LEHMANN, E.L. (1951). Some applications of the Cramér-Rao inequality. In *Proc. Second Berkeley Symp. on Math. Statist. Probab.*, Vol. 1. (J. Neyman, ed.), 13–21. University of California Press, Berkeley and Los Angeles, California, USA.

HOEFFDING, W. (1983). Unbiased range-preserving estimators. In *A Festschrift for Erich L. Lehmann.* (P.J. Bickel, K. A. Doksum and J. L. Hodges Jr., eds.), 249–260. Wadsworth International Group, Belmont, California, USA.

HOFERKAMP, C. and PEDDADA, S.D. (2002). Parameter estimation in linear models with heteroscedastic variances subject to order restrictions. J. Multivariate Anal., 82, 65–87.

*HOFFMANN, K. (1977). Admissibility of linear estimators with respect to restricted parameter sets. Math. Operationsforsch. Statist., Ser. Statistics, 8, 425–438.

*HOFFMANN, K. (1979). Characterization of minimax linear estimators in linear regression. Math. Operationsforsch. Statist., Ser. Statistics, 10, 19–26.

*HOFFMANN, K. (1980). Admissible improvements of the least squares estimator. Math. Operationsforsch. Statist., Ser. Statistics, 11, 373–388.

*HOFFMANN, K. (1995). All admissible linear estimators of the regression parameter vector in the case of an arbitrary parameter subset. J. Statist. Plann. Inference, 48, 371–377.

*HOFFMANN, K. (1996). A subclass of Bayes linear estimators that are minimax. Acta Appl. Math., 43, 87–95.

HU, F. (1994). Relevance weighted smoothing and a new bootstrap method, PhD thesis, Department of Statistics, The University of British Columbia, Vancouver, Canada.

HU, F. (1997). The asymptotic properties of maximum-relevance weighted likelihood estimations. Canad. J. Statist., 25, 45–59.

HU, F. and ZIDEK, J.V. (2001). The relevance weighted likelihood with applications. In Empirical Bayes and Likelihood Inference (S.E. Ahmed and N. Reid, eds.), 211–235. Springer Verlag.

HU, F. and ZIDEK, J.V. (2002). The weighted likelihood. Canad. J. Statist., 30, 347–371.

HU, X. (1997). Maximum-likelihood estimation under bound restriction and order and uniform bound restrictions. Statist. Probab. Lett., 35, 165–171.

HWANG, J.T. (1985). Universal domination and stochastic domination: estimation simultaneously under a broad class of loss functions. Ann. Statist., 13, 295–314.

HWANG, J.T. and PEDDADA, S.D. (1993). Confidence interval estimation under some restrictions on the parameters with non-linear boundaries. Statist. Probab. Lett., 18, 397–403.

HWANG, J.T. and PEDDADA, S.D. (1994). Confidence interval estimation subject to order restrictions. Ann. Statist., 22, 67–93.

ILIOPOULOS, G. (2000). A note on decision theoretic estimation of ordered parameters. Statist. Probab. Lett., 50, 33–38.

IWASA, M. and MORITANI, Y. (1997). A note on admissibility of the maximum likelihood estimator for a bounded normal mean. Statist. Probab. Lett, 32, 99–105.

IWASA, M. and MORITANI, Y. (2002). Concentration probabilities for restricted and unrestricted MLEs. J. Multivariate Anal., 80, 58–66.

JAFARI JOZANI, M., NEMATOLLAHI, N. and SHAFIE, K. (2002). An admissible minimax estimator of a bounded scale-parameter in a subclass of the exponential family under scale-invariant squared-error loss. Statist. Probab. Lett., 60, 437–444.

JAMES, W. and STEIN, C. (1961). Estimation with quadratic loss. In *Proc. Fourth Berkeley Symp. Math. Statist. Probab.*, Vol. 1 (J. Neyman, ed.), 311–319. University of California Press, Berkeley and Los Angeles. California, USA.

JEWEL, N.P. and KALBFLEISCH, J.D. (2004). Maximum likelihood estimation of ordered multinomial parameters. Biostatist., 5, 291–306.

JIN, C. and PAL, N. (1991). A note on the location parameters of two exponential distributions under order restrictions. Comm. Statist. Theory Methods, 20, 3147–3158.

JOHNSTONE, I.M. and MacGIBBON, K.B. (1992). Minimax estimation of a constrained Poisson vector. Ann. Statist., 20, 807–831.

JOOREL, SINGH J.P. and HOODA, B.K. (2002). On the estimation of ordered parameters of two uniform distributions. Aligarh J. Statist., 22, 101–108. Correction note: Aligarh J. Statist., 2005, submitted.

JUDGE, G.G., YANCEY, T.A. (1981). Sampling properties of an inequality restricted estimator. Econom. Lett., 7, 327–333.

JUDGE, G.G., YANCEY, T.A., BOCK, M.E. and BOHRER, R. (1984). The non-optimality of the inequality restricted estimator under squared error loss. J. Econom., 25, 165–177.

KAŁUSZKA, M. (1986). Admissible and minimax estimators of λ^r in the gamma distribution with truncated parameter space. Metrika, 33, 363–375.

KAŁUSZKA, M. (1988). Minimax estimation of a class of functions of the scale parameter in the gamma and other distributions in the case of truncated parameter space. Zastos. Mat., 20, 29–46.

KARLIN, S. (1957). Pólya type distributions, II. Ann. Math. Statist., 28, 281–308.

KATZ, M.W. (1961). Admissible and minimax estimates of parameters in truncated spaces. Ann. Math. Statist., 32, 136–142.

KATZ, M.W. (1963). Estimating ordered probabilities. Ann. Math. Statist., 34, 967–972.

KAUR, A. and SINGH, H. (1991). On the estimation of ordered means of two exponential populations. Ann. Inst. Statist. Math., 43, 347–356.

KEATING, J.P., MASON, R.L. and SEN, P.K. (1993). *Pitman's Measure of Closeness*, SIAM, Philadelphia, Pennsylvania, USA.

KELLY, R.E. (1989). Stochastic reduction of loss in estimating normal means by isotonic regression. Ann. Statist., 17, 937–940.

KEMPTHORNE, P.J. (1987). Numerical specification of discrete least favorable prior distributions. SIAM J. Sci. Statist. Comput., 8, 171–184.

KLAASSEN, C.A.J. (1989). The asymptotic spread of estimators. J. Statist. Plann. Inference, 23, 267–285.

KLEMM, R.J. and SPOSITO, V.A. (1980). Least squares solutions over interval restrictions. Comm. Statist. Simulation Comput., 9, 423–425.

KLOTZ, J.H., MILTON, R.C. and ZACKS, S. (1969). Mean square efficiency of estimators of variance components. J. Amer. Statist. Assoc., 64, 1383–1402.

*KNAUTZ, H. (1996). Linear plus quadratic (LPQ) quasiminimax estimation in the linear regression model. Acta Appl. Math., 43, 97–111.

KOUROUKLIS, S. (2000). Estimating the smallest scale parameter: Universal domination results. In *Probability and Statistical Models with Applications* (C.A. Charalambides, M.V. Koutras, N. Balahrishnan, eds.), Chapman and Hall/CRC.

*KOZÁK, J. (1985). Modified minimax estimation of regression coefficients. Statistics, 16, 363–371.

*KUBÁČEK, L. (1995). Linear statistical models with constraints revisited. Math. Slovaca, 45, 287–307.

KUBOKAWA, T. (1994a). Double shrinkage estimation of ratio of scale parameters. Ann. Inst. Statist. Math., 46, 95–116.

KUBOKAWA, T. (1994b). A unified approach to improving equivariant estimators, Ann. Statist.,22, 290–299.

KUBOKAWA, T. and SALEH, A.K.MD.E. (1994). Estimation of location and scale parameters under order restrictions. J. Statist. Res., 28, 41–51.

KUMAR, S. and SHARMA, D. (1988). Simultaneous estimation of ordered parameters. Comm. Statist. Theory Methods, 17, 4315–4336.

KUMAR, S. and SHARMA, D. (1989). On the Pitman estimator of ordered normal means. Comm. Statist. Theory Methods, 18, 4163–4175.

KUMAR, S. and SHARMA, D. (1992). An inadmissibility result for affine equivariant estimators. Statist. Decisions, 10, 87–97.

KUMAR, S. and SHARMA, D. (1993). Minimaxity of the Pitman estimator of ordered normal means when the variances are unequal. J. Indian Soc. Agricultural Statist., 45, 230–234.

KURIKI, S. and TAKEMURA, A. (2000). Shrinkage estimation towards a closed convex set with a smooth boundary. J. Multivariate Anal., 75, 79–111.

KUSHARY, D. and COHEN, A. (1989). Estimating ordered location and scale parameters. Statist. Decisions, 7, 201–213.

KUSHARY, D. and COHEN, A. (1991). Estimation of ordered Poisson parameters. Sankhyā Ser. A, 53, 334–356.

*LAMOTTE, L.R. (1982). Admissibility in linear estimation. Ann. Statist., 10, 245–255.

*LAMOTTE, L.R. (1997). On limits of uniquely best linear estimators. Metrika, 45, 197–211.

*LÄUTER, H. (1975). A minimax linear estimator for linear parameters under restrictions in form of inequalities. Math. Operationsforsch. u. Statist., 6, 689–695.

*LAUTERBACH, J. and STAHLECKER, P. (1988). Approximate minimax estimation in linear regression : a simulation study. Comm. Statist. Simulation Comput., 17, 209–227.

*LAUTERBACH, J. and STAHLECKER, P. (1990). Some properties of $[\mathrm{tr}(Q^{2p})]^{1/2p}$ with application to linear minimax estimation. Linear Algebra and Appl., 127, 301–325.

*LAUTERBACH, J. and STAHLECKER, P. (1992). A numerical method for an approximate minimax estimator in linear regression. Linear Algebra Appl., 176, 91–108.

LEE, C.C. (1981). The quadratic loss of isotonic regression under normality. Ann. Statist., 9, 686–688.

LEE, C.C. (1983). The min-max algorithm and isotonic regression. Ann. Statist., 11, 467–477.

LEE, C.C. (1988). The quadratic loss of order restricted estimators for treatment means with a control. Ann. Statist., 16, 751–758.

LEHMANN, E.L. (1983). Theory of Point Estimation. John Wiley & Sons.

LEHMANN, E.L. and CASELLA, G. (1998). Theory of Point Estimation. Second edition, Springer Verlag.

LILLO, R.E. and MARTÍN, M. (2000). Bayesian approach to estimation of ordered uniform scale parameters. J. Statist. Plann. Inference, 87, 105–118.

LOVELL, M.C. and PRESCOTT, E. (1970). Multiple regression with inequality constraints: pretesting bias, hypothesis testing and efficiency. J. Amer. Statist. Assoc., 65, 913–925.

MARCHAND, É and MacGIBBON, B. (2000). Minimax estimation of a constrained binomial proportion. Statist. Decisions, 18, 129–167.

MARCHAND, É. and PERRON, F. (2001). Improving on the MLE of a bounded normal mean. Ann. Statist., 29, 1078–1093.

MARCHAND, É. and PERRON, F. (2002). On the minimax estimator of a bounded normal mean. Statist. Probab. Lett., 58, 327–333.

MARCHAND, É. and PERRON, F. (2005). Improving on the mle of a bounded location parameter for spherical distributions. J. Multivariate Anal., 92, 227–238.

MARUYAMA, Y. and IWASAKI, K. (2005). Sensitivity of minimaxity and admissibility in the estimation of a positive normal mean. Ann. Inst. Statist. Math., 57, 145–156.

*MATHEW, T. (1985). Admissible linear estimation in singular linear models with respect to a restricted parameter set. Comm. Statist. Theory Methods, 14, 491–498.

MATHEW, T., SINHA, B.K. and SUTRADHAR, B.C. (1992). Nonnegative estimation of variance components in unbalanced mixed models with two variance components. J. Multivariate Anal., 42, 77–101.

MEHTA, J.S. and GURLAND, J. (1969). Combinations of unbiased estimates of the mean which consider inequality of unknown variances. J. Amer. Statist. Assoc., 64, 1042–1055.

MENÉNDEZ, J.A. and SALVADOR, B. (1987). An algorithm for isotonic median regression. Comput. Statist. Data Anal., 5, 399–406. Correction: Comput. Statist. Data Anal., 11, 1991, 203–204.

MISRA, N., CHOUDHARY, P.K., DHARIYAL, I.D. and KUNDU, D. (2002). Smooth estimators for estimating order restricted scale parameters of two gamma distributions. Metrika, 56, 143–161.

MISRA, N. and DHARIYAL, I.D. (1995). Some inadmissibility results for estimating ordered uniform scale parameters. Comm. Statist. Theory Methods, 24, 675–685.

MISRA, N. and SINGH, H. (1994). Estimation of ordered location parameters : the exponential distribution. Statistics, 25, 239–249.

MISRA, N. and VAN DER MEULEN, E.C. (1997). On estimation of the common mean of k (≥ 2) normal populations with order restricted variances. Statist. Probab. Lett., 36, 261–267.

MISRA, N. and VAN DER MEULEN, E.C. (2005). On estimation of the common mean of $k(\geq 2)$ normal populations when the ordering between some (or all) of the variances is known. (Unpublished manuscript)

MOORS, J.J.A. (1981). Inadmissibility of linearly invariant estimators in truncated parameter spaces. J. Amer. Statist. Assoc., 76, 910–915.

MOORS, J.J.A. (1985). Estimation in Truncated Parameter Spaces. Ph.D thesis, Tilburg University, Tilburg, The Netherlands.

MOORS, J.J.A. and VAN HOUWELINGEN, J.C. (1993). Estimation of linear models with inequality restrictions. Statist. Neerlandica, 47, 185–198.

MUDHOLKAR, G.S., SUBBBAIAH, P. and GEORGE, S. (1977). A note on the estimation of two ordered Poisson parameters. Metrika, 24, (1977), 87–98.

OHTANI, K. (1987). The MSE of the least squares estimators over an interval constraint. Econom. Lett., 25, 351–354.

OHTANI, K. (1991). Small sample properties of the interval constrained least squares estimator when the error terms have a multivariate t distribution. J. Japan Statist. Soc., 21, 197–204.

OHTANI, K. (1996). On an adjustment of degrees of freedom in the minimum mean squared error estimator. Comm. Statist. Theory Methods, 25, 3049–3058.

OHTANI, K. and WAN, A.T.K. (1998). On the sampling performance of an improved Stein inequality restricted estimator. Aust. N.Z. J. Statist., 40, 181–187.

OUASSOU, I. and STRAWDERMAN, W.E. (2002). Estimation of a parameter vector restricted to a cone. Statist. Probab. Lett., 56, 121–129.

PAL, N. and KUSHARY, D. (1992). On order restricted location parameters of two exponential distributions. Statist. Decisions, 10, 133–152.

PARK, C.G. (1998). Least squares estimation of two functions under order restriction in isotonicity. Statist. Probab. Lett., 37, 97–100.

PARSIAN, A. and NEMATOLLAHI, N. (1995). On the admissibility of ordered Poison parameter under the entropy loss function. Comm. Statist. Theory Methods, 24, 2451–2467.

PARSIAN, A. and SANJARI FARSIPOUR, N. (1997). Estimation of parameters of exponential distribution in the truncated space using asymmetric loss function. Statist. Papers, 38, 423–443.

PARSIAN, A., SANJARI FARSIPOUR, N. and NEMATOLLAHI, N. (1993). On the minimaxity of Pitman type estimator under a linex loss function. Comm. Statist. Theory Methods, 22, 97–113.

PERRON, F. (2003). Improving on the MLE of p for a binomial(n, p) when p is around $1/2$. In *Mathematical Statistics and Applications: Festschrift for Constance van Eeden* (M. Moore, S. Froda and C. Leger, eds.), IMS Lecture Notes and Monograph Series, 43, 45–61. Institute of Mathematical Statistics, Hayward, California, USA.

*PILZ, J. (1986). Minimax linear regression estimation with symmetric parameter restrictions. J. Statist. Plann. Inference, 13, 297–318.

PITMAN, E.J.G. (1937). The "closest" estimates of statistical parameters. Proc. Cambridge Philos. Soc., 33, 212–222.

PITMAN, E.J.G. (1939). The estimation of the location and scale parameters of a continuous population of any given form. Biometrika, 30, 391–421.

PURI, P.S. and SINGH, H. (1990). On recursive formulas for isotonic regression useful for statistical inference under order restrictions. J. Statist. Plann. Inference, 24, 1–11.

QIAN, S. (1992). Minimum lower sets algorithms for isotonic regression. Statist. Probab. Lett., 15, 31–35.

QIAN, S. (1994a). The structure of isotonic regression class for LAD problems with quasi-order constraints. Comput. Statist. Data Anal., 18, 389–401.

QIAN, S. (1994b). Generalization of least-square isotonic regression. J. Statist. Plann. Inference, 38, 389–397.

QIAN, S. (1996). An algorithm for tree-ordered isotonic median regression. Statist. Probab. Lett., 27, 195–199.

RAHMAN, M.S. and GUPTA, R.P. (1993). Family of transformed chi-square distributions. Comm. Statist. Theory Methods, 22, 135–146.

ROBERT, C.P. (1997), *The Bayesian Choice. A Decision Theoretic Motivation.* Corrected third printing. Springer-Verlag.

ROBERTSON, T. and WALTMAN, P. (1968). On estimating monotone parameters. Ann. Math. Statist., 39, 1030–1039.

ROBERTSON, T. and WRIGHT, F.T. (1973). Multiple isotonic median regression. Ann. Statist., 1, 422–432.

ROBERTSON, T. and WRIGHT, F.T. (1980). Algorithms in order restricted statistical inference and the Cauchy mean value property. Ann. Statist., 8, 645–651.

ROBERTSON, T., WRIGHT, F.T. and DYKSTRA, R.L. (1988). *Order Restricted Statistical Inference.* John Wiley & Sons.

RUEDA, C. and SALVADOR, B. (1995). Reduction of risk using restricted estimators. Comm. Statist. Theory Methods, 24, 1011–1022.

RUEDA, C., SALVADOR, B. and FERNÁNDEZ, M.A. (1997a). Simultaneous estimation in a restricted linear model. J. Multivariate Anal., 61, 61–66.

RUEDA, C., SALVADOR, B. and FERNÁNDEZ, M.A. (1997b). A good property of the maximum likelihood estimator in a restricted normal model. Test, 6, 127–135.

RUYMGAART, F.H. (1996). The van Trees inequality derived from the Cramér-Rao inequality and applied to nonparametric regression. In *Research Developments in Statistics and Probability,* Festschrift in honor of Madan Puri on the occasion of his 65th birthday (E. Brunner and M. Denker, eds.), 219–230, VSP.

SACKROWITZ, H. (1982). Procedures for improving the MLE for ordered binomial parameters. J. Statist. Plann. Inference, 6, 287–296.

SACKROWITZ, H. and STRAWDERMAN, W. (1974). On the admissibility of the M.L.E. for ordered binomial parameters. Ann. Statist., 2, 822–828.

SACKS, J. (1960). Generalized Bayes solutions in estimation problems. Abstract 75, Ann. Math. Statist., 31, 246.

SACKS, J. (1963). Generalized Bayes solutions in estimation problems. Ann. Math. Statist., 34, 751–768.

SAMPSON, A.R., SINGH, H. and WHITAKER, L.R. (2003). Order restricted estimators: some bias results. Statist. Probab. Lett., 61, 299–308. Correction note, Statist. Probab. Lett., submitted (2006).

SANJARI FARSIPOUR, N. (2002). Pitman nearness comparison of estimators of parameters of exponential distribution in the truncated space. J. Appl. Statist. Sci., 11, 187–194.

SASABUCHI, S., INUTSUKA, M. and KULATUNGA. D.D.S. (1983). A multivariate version of isotonic regression. Biometrika, 70, 465–472.

SATO, M. and AKAHIRA, M. (1995). Information inequalities for the minimax risk. J. Japan Statist. Soc., 25, 151–158.

SATO, M. and AKAHIRA, M. (1996). An information inequality for the Bayes risk. Ann. Statist., 24, 2288–2295.

*SCHIPP, B. (1993). Approximate minimax estimators in the simultaneous equations model. J. Statist. Plann. Inference, 36, 197–214.

*SCHIPP, B. and TOUTENBURG, H. (1996). Feasible minimax estimators in the simultaneous equations model under partial restrictions. J. Statist. Plann. Inference, 50, 241–250.

*SCHIPP, B., TRENKLER, G. and STAHLECKER, P. (1988). Minimax estimation with additional linear restrictions - a simulation study. Comm. Statist. Simulation Comput., 17, 393–406.

SENGUPTA, D. and SEN, P.K. (1991). Shrinkage estimation in a restricted parameter space. Sankhyā Ser. A, 53, 389–411.

SHAO, P.Y-S. and STRAWDERMAN, W. E. (1994). Improving on truncated estimators. In *Statistical Decision Theory and Related Topics V* (S.S. Gupta and J.O. Berger, eds.), 369–376. Springer-Verlag.

SHAO, P.Y-S. and STRAWDERMAN, W.E. (1996a). Improving on truncated linear estimates of exponential and gamma scale parameters. Canad. J. Statist., 24, 105–114.

SHAO, P.Y-S. and STRAWDERMAN, W.E. (1996b). Improving on the MLE of a positive normal mean. Statist. Sinica, 6, 259–274.

SHINOZAKI, N. and CHANG, Y.-T. (1999). A comparison of maximum likelihood and best unbiased estimators in the estimation of linear combinations of positive normal means. Statist. Decisions, 17, 125–136.

SINGH, H., GUPTA, R.D. and MISRA, N. (1993). Estimation of parameters of an exponential distribution when the parameter space is restricted with an application to two-sample problem. Comm. Statist. Theory Methods, 22, 461–477.

SINHA, B.K. (1979). Is the maximum likelihood estimate of the common mean of several normal population admissible? Sankhyā Ser. B, 40, 192–196.

*SRIVASTAVA, A.K. and SHUKLA, P. (1996). Minimax estimation in linear regression model. J. Statist. Plann. Inference, 50, 77–89.

*STAHLECKER, P., JÄNNER, M. and SCHMIDT, K. (1991). Linear-affine Minimax-Schätzer unter Ungleichungsrestriktionen. Allgemein. Statist. Arch., 75, 245–264.

STAHLECKER, P., KNAUTZ, H. and TRENKLER, G. (1996). Minimax adjustment technique in a parameter restricted linear model. Acta Appl. Math., 43, 139–144.

*STAHLECKER, P. and LAUTERBACH, J. (1987). Approximate minimax estimation in linear regression : theoretical results. Comm. Statist. Theory Methods, 16, 1101–1116.

*STAHLECKER, P. and LAUTERBACH, J. (1989). Approximate linear minimax estimation in regression analysis with ellipsoidal constraints. Comm. Statist. Theory Methods, 18, 2755–2784.

*STAHLECKER, P. and SCHMIDT, K. (1989). Approximation linearer Ungleichungsrestriktionen im linearen Regressionsmodell. Allgemein. Statist. Arch., 73, 184–194.

*STAHLECKER, P. and TRENKLER, G. (1988). Full and partial minimax estimation in regression analysis with additional linear constraints, Linear Algebra Appl., 111, 279–292.

*STAHLECKER, P. and TRENKLER, G. (1993). Minimax estimation in linear regression with singular covariance structure and convex polyhedral constraints. J. Statist. Plann. Inference, 36, 185–196.

STEIN, C. (1956). Inadmissibility of the usual estimator for the mean of a multivariate normal distribution. In Proc. Third Berkeley Symp. Math. Statist. Probab., Vol. 1. (J. Neyman ed.), 197–206. University of California Press, Berkeley and Los Angeles, California, USA.

STEIN, C. (1964). Inadmissibility of the usual estimator for the variance of a normal distribution with unknown mean. Ann. Inst. Statist. Math., 16, 155–160.

STRAWDERMAN, W.E. (1974). Minimax estimation of powers of the variance of a normal population under squared error loss. Ann. Statist., 2, 190–198.

STRAWDERMAN, W.E. (1999). Personal communication.

STRÖMBERG, U. (1991). An algorithm for isotonic regression with arbitrary convex distance function. Comput. Statist. Data Anal., 11, 205–219.

*TERÄSVIRTA, T. (1989). Estimating linear models with incomplete ellipsoidal restrictions. Statistics, 20, 187–194.

THOMPSON, W.A. (1962). The problem of negative estimates of variance components. Ann. Math. Statist., 33, 273–289.

THOMSON, M. (1982). Some results on the statistical properties of an inequality constrained least squares estimator in a linear model with two regressors. J. Econometrics , 19, 215–231.

THOMSON, M. and SCHMIDT, P. (1982). A note on the comparison of the mean square error of inequality constrained least squares and other related estimators. Rev. Econom. Stat., 64, 174–176.

*TOUTENBURG, H. (1980). On the combination of equality and inequality restrictions on regression coefficients. Biometrical J., 22, 271–274.

*TOUTENBURG, H. and ROEDER, B. (1978). Minimax-linear and Theil estimator for restrained regression coefficients. Math. Operationsforsch. Statist., Ser. Statistics, 9, 499–505.

*TOUTENBURG, H. and SRIVASTAVA, V.K. (1996). Estimation of regression coefficients subject to interval constraints. Sankhyā Ser. A, 58, 273–282.

TOWHIDI, M. and BEHBOODIAN, J. (2002). Minimax estimation of a bounded parameter under some bounded loss functions. Far East J. Theo. Statist., 6, 39–48.

VAN EEDEN, C. (1956). Maximum likelihood estimation of ordered probabilities. Proc. Kon. Nederl. Akad. Wetensch. Ser. A, 59, 444–455.

VAN EEDEN, C, (1957a). Maximum likelihood estimation of partially or completely ordered parameters. Proc. Kon. Nederl. Akad. Wetensch. Ser. A, 60, 128–136 and 201–211.

VAN EEDEN, C. (1957b). Note on two methods for estimating ordered parameters of probability distributions. Proc. Kon. Nederl. Akad. Wetensch. Ser. A, 60, 506–512.

VAN EEDEN, C. (1957c). A least-squares inequality for maximum likelihood estimates of ordered parameters. Proc. Kon. Nederl. Akad. Wetensch. Ser. A, 60, 513–521.

VAN EEDEN, C. (1958). *Testing and Estimating Ordered Parameters of Probability distributions.* PhD. thesis, University of Amsterdam, Amsterdam, The Netherlands.

VAN EEDEN, C. (1995). Minimax estimation of a lower-bounded scale parameter of a gamma distribution for scale-invariant squared-error loss. Canad. J. Statist., 23, 245–256.

VAN EEDEN, C. (2000). Minimax estimation of a lower-bounded scale parameter of an F-distribution. Statist. Probab. Lett., 46, 283–286.

VAN EEDEN, C. and ZIDEK, J.V. (1994a). Group Bayes estimation of the exponential mean : A retrospective view of the Wald theory. In *Statistical Decision Theory and Related Topics V* (S. S. Gupta and J. O. Berger, eds.), 35–49. Springer-Verlag.

VAN EEDEN, C. and ZIDEK, J.V. (1994b). Group-Bayes estimation of the exponential mean: A preposterior analysis. Test, 3, 125–143; corrections p. 247.

VAN EEDEN, C. and ZIDEK, J.V. (1999). Minimax estimation of a bounded scale parameter for scale-invariant squared-error loss. Statist. Decisions, 17, 1–30.

VAN EEDEN, C. and ZIDEK, J.V. (2001). Estimating one of two normal means when their difference is bounded. Statist. Probab. Lett., 51, 277–284.

VAN EEDEN, C. and ZIDEK, J.V. (2002). Combining sample information in estimating ordered normal means. Sankhyā Ser. A, 64, 588–610.

VAN EEDEN, C. and ZIDEK, J.V. (2004). Combining the data from two normal populations to estimate the mean of one when their means difference is bounded. J. Multvariate Anal., 88, 19–46.

VAN TREES, H.L. (1968). *Detection, Estimation and Modulation Theory*, Part 1. John Wiley.

VIDAKOVIC, B. and DASGUPTA, A. (1995). Lower bounds on Bayes risks for estimating a normal variance: With applications. Canad. J. Statist., 23, 269–282.

VIDAKOVIC, B. and DASGUPTA, A. (1996). Efficiency of linear rules for estimating a bounded normal mean. Sankyā Ser. A., 58, 81–100.

VIJAYASREE, G., MISRA, N. and SINGH, H. (1995). Componentwise estimation of ordered parameters of k (≥ 2) exponential populations. Ann. Inst. Statist. Math., 47, 287–307.

VIJAYASREE, G. and SINGH, H. (1991). Simultaneous estimation of two ordered exponential parameters. Comm. Statist. Theory Methods, 20, 2559–2576.

VIJAYASREE, G. and SINGH, H. (1993). Mixed estimators of two ordered exponential means. J. Statist. Plann. Inference, 35, 47–53.

WALD, A. (1950). *Statistical Decision Functions*, John Wiley.

WAN, A.T.K. (1994a). The non-optimality of interval restricted and pre-test estimators under squared error loss. Comm. Statist. Theory Methods, 23, 2231–2252.

WAN, A.T.K. (1994b). Risk comparison of the inequality constrained least squares and other related estimators under balanced loss. Econom. Lett., 46, 203–210.

WAN, A.T.K. and OHTANI, K. (2000). Minimum mean-squared error estimation in linear regression with an inequality constraint. J. Statist. Plann. Inference, 86, 157–173.

WAN, A.T.K., ZOU, G. and LEE, A.H. (2000). Minimax and Γ-minimax estimation for the Poisson distribution under LINEX loss when the parameter space is restricted. Statist. Probab. Lett., 50, 23–32.

WANG, S. (2001). The maximum weighted likelihood estimator. PhD thesis, Department of Statistics, The University of British Columbia, Vancouver, Canada.

ZEYTINOGLU, M. and MINTZ, M. (1984). Optimal fixed size confidence procedures for a restricted parameter space. Ann. Statist., 12, 945–957.

ZINZIUS, E. (1979). *Beitrage zur Theorie der Nichtsequentiellen Parameterschätzprobleme*. PhD thesis, University of Karlsruhe, Karlsruhe, Germany.

ZINZIUS, E. (1981). Minimaxschätzer für den Mittelwert ϑ einer normalverteilten Zufallsgröße mit bekannter Varianz bei vorgegebener oberer und unterer Schranke für ϑ. Math. Operationsforsch. Statist., Ser. Statistics, 12, 551–557.

ZOU, G.-H. (1993). Minimax estimation in a restricted parameter space. J. Systems Sci. Math. Sci., 13, 345–348.

ZUBRZYCKI, S. (1966). Explicit formulas for minimax admissible estimators in some cases of restrictions imposed on the parameter. Zastos. Mat., 9, 31–52.

Author Index

Subject Index

Lecture Notes in Statistics

For information about Volumes 1 to 133,
please contact Springer-Verlag

134: Yu. A. Kutoyants, Statistical Inference
For Spatial Poisson Processes. vii, 284 pp.,
1998.

135: Christian P. Robert, Discretization and
MCMC Convergence Assessment. x, 192
pp., 1998.

136: Gregory C. Reinsel, Raja P. Velu,
Multivariate Reduced-Rank Regression. xiii,
272 pp., 1998.

137: V. Seshadri, The Inverse Gaussian
Distribution: Statistical Theory and
Applications. xii, 360 pp., 1998.

138: Peter Hellekalek and Gerhard Larcher
(Editors), Random and Quasi-Random
Point Sets. xi, 352 pp., 1998.

139: Roger B. Nelsen, An Introduction to
Copulas. xi, 232 pp., 1999.

140: Constantine Gatsonis, Robert E. Kass,
Bradley Carlin, Alicia Carriquiry, Andrew
Gelman, Isabella Verdinelli, and Mike West
(Editors), Case Studies in Bayesian
Statistics, Volume IV. xvi, 456 pp., 1999.

141: Peter Müller and Brani Vidakovic
(Editors), Bayesian Inference in Wavelet
Based Models. xiii, 394 pp., 1999.

142: György Terdik, Bilinear Stochastic
Models and Related Problems of Nonlinear
Time Series Analysis: A Frequency Domain
Approach. xi, 258 pp., 1999.

143: Russell Barton, Graphical Methods
for the Design of Experiments. x, 208 pp.,
1999.

144: L. Mark Berliner, Douglas Nychka,
and Timothy Hoar (Editors), Case Studies
in Statistics and the Atmospheric Sciences.
x, 208 pp., 2000.

145: James H. Matis and Thomas R. Kiffe,
Stochastic Population Models. viii, 220 pp.,
2000.

146: Wim Schoutens, Stochastic Processes
and Orthogonal Polynomials. xiv, 163 pp.,
2000.

147: Jürgen Franke, Wolfgang Härdle, and
Gerhard Stahl, Measuring Risk in Complex
Stochastic Systems. xvi, 272 pp., 2000.

148: S.E. Ahmed and Nancy Reid, Empirical
Bayes and Likelihood Inference. x, 200 pp.,
2000.

149: D. Bosq, Linear Processes in Function
Spaces: Theory and Applications. xv, 296
pp., 2000.

150: Tadeusz Caliński and Sanpei
Kageyama, Block Designs: A
Randomization Approach, Volume I:
Analysis. ix, 313 pp., 2000.

151: Håkan Andersson and Tom Britton,
Stochastic Epidemic Models and Their
Statistical Analysis. ix, 152 pp., 2000.

152: David Rios Insua and Fabrizio Ruggeri,
Robust Bayesian Analysis. xiii, 435 pp., 2000.

153: Parimal Mukhopadhyay, Topics in
Survey Sampling. x, 303 pp., 2000.

154: Regina Kaiser and Agustín Maravall,
Measuring Business Cycles in Economic
Time Series. vi, 190 pp., 2000.

155: Leon Willenborg and Ton de Waal,
Elements of Statistical Disclosure Control.
xvii, 289 pp., 2000.

156: Gordon Willmot and X. Sheldon Lin,
Lundberg Approximations for Compound
Distributions with Insurance Applications.
xi, 272 pp., 2000.

157: Anne Boomsma, Marijtje A.J. van
Duijn, and Tom A.B. Snijders (Editors),
Essays on Item Response Theory. xv, 448
pp., 2000.

158: Dominique Ladiray and Benoît
Quenneville, Seasonal Adjustment with the
X-11 Method. xxii, 220 pp., 2001.

159: Marc Moore (Editor), Spatial Statistics:
Methodological Aspects and Some
Applications. xvi, 282 pp., 2001.

160: Tomasz Rychlik, Projecting Statistical
Functionals. viii, 184 pp., 2001.

161: Maarten Jansen, Noise Reduction by
Wavelet Thresholding. xxii, 224 pp., 2001.

162: Constantine Gatsonis, Bradley Carlin, Alicia Carriquiry, Andrew Gelman, Robert E. Kass Isabella Verdinelli, and Mike West (Editors), Case Studies in Bayesian Statistics, Volume V. xiv, 448 pp., 2001.

163: Erkki P. Liski, Nripes K. Mandal, Kirti R. Shah, and Bikas K. Sinha, Topics in Optimal Design. xii, 164 pp., 2002.

164: Peter Goos, The Optimal Design of Blocked and Split-Plot Experiments. xiv, 244 pp., 2002.

165: Karl Mosler, Multivariate Dispersion, Central Regions and Depth: The Lift Zonoid Approach. xii, 280 pp., 2002.

166: Hira L. Koul, Weighted Empirical Processes in Dynamic Nonlinear Models, Second Edition. xiii, 425 pp., 2002.

167: Constantine Gatsonis, Alicia Carriquiry, Andrew Gelman, David Higdon, Robert E. Kass, Donna Pauler, and Isabella Verdinelli (Editors), Case Studies in Bayesian Statistics, Volume VI. xiv, 376 pp., 2002.

168: Susanne Rässler, Statistical Matching: A Frequentist Theory, Practical Applications and Alternative Bayesian Approaches. xviii, 238 pp., 2002.

169: Yu. I. Ingster and Irina A. Suslina, Nonparametric Goodness-of-Fit Testing Under Gaussian Models. xiv, 453 pp., 2003.

170: Tadeusz Caliński and Sanpei Kageyama, Block Designs: A Randomization Approach, Volume II: Design. xii, 351 pp., 2003.

171: D.D. Denison, M.H. Hansen, C.C. Holmes, B. Mallick, B. Yu (Editors), Nonlinear Estimation and Classification. x, 474 pp., 2002.

172: Sneh Gulati, William J. Padgett, Parametric and Nonparametric Inference from Record-Breaking Data. ix, 112 pp., 2002.

173: Jesper Møller (Editors), Spatial Statistics and Computational Methods. xi, 214 pp., 2002.

174: Yasuko Chikuse, Statistics on Special Manifolds. xi, 418 pp., 2002.

175: Jürgen Gross, Linear Regression. xiv, 394 pp., 2003.

176: Zehua Chen, Zhidong Bai, Bimal K. Sinha, Ranked Set Sampling: Theory and Applications. xii, 224 pp., 2003.

177: Caitlin Buck and Andrew Millard (Editors), Tools for Constructing Chronologies: Crossing Disciplinary Boundaries, xvi, 263 pp., 2004.

178: Gauri Sankar Datta and Rahul Mukerjee, Probability Matching Priors: Higher Order Asymptotics, x, 144 pp., 2004.

179: D.Y. Lin and P.J. Heagerty (Editors), Proceedings of the Second Seattle Symposium in Biostatistics: Analysis of Correlated Data, vii, 336 pp., 2004.

180: Yanhong Wu, Inference for Change-Point and Post-Change Means After a CUSUM Test, xiv, 176 pp., 2004.

181: Daniel Straumann, Estimation in Conditionally Heteroscedastic Time Series Models, x, 250 pp., 2004.

182: Lixing Zhu, Nonparametric Monte Carlo Tests and Their Applications, xi, 192 pp., 2005.

183: Michel Bilodeau, Fernand Meyer, and Michel Schmitt (Editors), Space, Structure and Randomness, xiv, 416 pp., 2005.

184: Viatcheslav B. Melas, Functional Approach to Optimal Experimental Design, vii., 352 pp., 2005.

185: Adrian Baddeley, Pablo Gregori, Jorge Mateu, Radu Stoica, and Dietrich Stoyan, (Editors), Case Studies in Spatial Point Process Modeling, xiii., 324 pp., 2005.

186: Estela Bee Dagum and Pierre A. Cholette, Benchmarking, Temporal Distribution, and Reconciliation Methods for Time Series, xiv., 410 pp., 2006.

187: Patrice Bertail, Paul Doukhan and Philippe Soulier, (Editors), Dependence in Probability and Statistics, viii., 504 pp., 2006.

188: Constance van Eeden, Restricted Parameter Space Estimation Problems, vi, 176 pp., 2006.

 **Springer**
the language of science

springer.com

 ## Dependence in Probability and Statistics

P. Bertail. P Doukhan, and P. Soulier (Editors)

This book gives a detailed account of some recent developments in the field of probability and statistics for dependent data. The book covers a wide range of topics from Markov chain theory and weak dependence with an emphasis on some recent developments on dynamical systems, to strong dependence in times series and random fields. A special section is devoted to statistical estimation problems and specific applications.

2006. 504 p. (Lecture Notes in Statistics, Vol. 187) Softcover
ISBN 0-387-31741-4

 ## Benchmarking, Temporal Distribution, and Reconciliation Methods for Time Series

Estela Bee Dagum and Pierre A. Cholette

This book discusses the statistical methods most often applied for adjustments, ranging from ad hoc procedures to regression-based models. The latter are emphasized, because of their clarity, ease of application, and superior results. Each topic is illustrated with many real case examples. In order to facilitate understanding of their properties and limitations of the methods discussed, a real data example, the Canada Total Retail Trade Series, is followed throughout the book.

2006. 410 p. (Lecture Notes in Statistics, Vol. 186) Softcover
ISBN 0-387-31102-5

 ## Case Studies in Spatial Point Process Modeling

A. Baddeley, P. Gregori, J. Mateu, R. Stoica and D. Stoyan (Editors)

Point process statistics is successfully used in fields such as material science, human epidemiology, social sciences, animal epidemiology, biology, and seismology. Its further application depends greatly on good software and instructive case studies that show the way to successful work. This book satisfies this need by a presentation of the spatstat package and many statistical examples.

2005. 312 p. (Lecture Notes in Statistics, Vol. 185) Softcover
ISBN 0-387-28311-0

Easy Ways to Order▶   Call: Toll-Free 1-800-SPRINGER • E-mail: orders-ny@springer.com • Write: Springer, Dept. S8113, PO Box 2485, Secaucus, NJ 07096-2485 • Visit: Your local scientific bookstore or urge your librarian to order.